On the Economy and Management of the Dairy

A Guide to the Methods and Equipment of Livestock Farming

By

William Youatt

Copyright © 2011 Read Books Ltd.
This book is copyright and may not be
reproduced or copied in any way without
the express permission of the publisher in writing

British Library Cataloguing-in-Publication Data
A catalogue record for this book is available from
the British Library

Farming

Agriculture, also called farming or husbandry, is the cultivation of animals, plants, or fungi for fibre, biofuel, drugs and other products used to sustain and enhance human life. Agriculture was the key development in the rise of sedentary human civilization, whereby farming of domesticated species created food surpluses that nurtured the development of civilization. It is hence, of extraordinary importance for the development of society, as we know it today. The word *agriculture* is a late Middle English adaptation of Latin *agricultūra*, from *ager*, 'field', and *cultūra*, 'cultivation' or 'growing'. The history of agriculture dates back thousands of years, and its development has been driven and defined by vastly different climates, cultures, and technologies. However all farming generally relies on techniques to expand and maintain the lands that are suitable for raising domesticated species. For plants, this usually requires some form of irrigation, although there are methods of dryland farming. Livestock are raised in a combination of grassland-based and landless systems, in an industry that covers almost one-third of the world's ice- and water-free area.

Agricultural practices such as irrigation, crop rotation, fertilizers, pesticides and the domestication of livestock were developed long ago, but have made great progress in the past century. The history of agriculture has played a major role in human history, as agricultural progress has been a crucial factor in worldwide socio-economic change. Division of labour in agricultural societies made (now) commonplace specializations, rarely seen in hunter-gatherer cultures, which allowed the growth of towns and cities, and the complex societies we call civilizations. When farmers became capable of producing food beyond the needs of their own families, others in their society were freed to devote themselves to projects other than food acquisition. Historians and anthropologists have long argued that the development of agriculture made civilization possible.

In the developed world, industrial agriculture based on large-scale monoculture has become the dominant system of modern farming, although there is growing support for sustainable agriculture, including permaculture and organic agriculture. Until the Industrial Revolution, the vast majority of the human population laboured in agriculture. Pre-industrial agriculture was typically for self-sustenance, in which farmers raised most of their crops for their own consumption, instead of cash crops for trade. A remarkable shift in agricultural practices has occurred over the past two centuries however, in response to new technologies, and the development of world markets. This also has led to technological improvements in agricultural techniques, such as the Haber-Bosch method for synthesizing ammonium nitrate which made the traditional practice of recycling nutrients with crop rotation and animal manure less important.

Modern agronomy, plant breeding, agrochemicals such as pesticides and fertilizers, and technological improvements have sharply increased yields from cultivation, but at the same time have caused widespread ecological damage and negative human health effects. Selective breeding and modern practices in animal husbandry have similarly increased the output of meat, but have raised concerns about animal welfare and the health effects of the antibiotics, growth hormones, and other chemicals commonly used in industrial meat production. Genetically Modified Organisms are an increasing component of agriculture today, although they are banned in several countries. Another controversial issue is 'water management'; an increasingly global issue fostering debate. Significant degradation of land and water resources, including the depletion of aquifers, has been observed in recent decades, and the effects of global warming on agriculture and of agriculture on global warming are still not fully understood.

The agricultural world of today is at a cross roads. Over one third of the worlds workers are employed in agriculture, second only to the services sector, but its future is uncertain. A constantly growing world population is necessitating more and more land being utilised for growth of food stuffs, but also the burgeoning mechanised methods of food cultivation and harvesting means that many farming jobs are becoming redundant. Quite how the sector will respond to these challenges remains to be seen.

Cattle Farming

Cattle are the most common type of large domesticated ungulates (a group of mammals, mainly categorised by their hoofs). They are a prominent modern member of the subfamily *Bovinae*, and are the most widespread species of the genus *Bos*. Cows, or cattle, are commonly raised as livestock for meat, as well as dairy animals and even draft animals, kept for such errands as pulling carts, plows and the like. Other products include leather and dung for manure or fuel. From as few as 80 progenitors domesticated in southeast Turkey and northern Iraq about 10,500 years ago, an estimated 1.3 billion cattle are in the world today. Cattle occupy a unique role in human history, they are one of the few animals to have been domesticated since at least the early Neolithic period – and they have been seen variously as workers, sacred animals, and foodstuffs.

Cattle are often raised by allowing herds to graze on the grasses of large tracts of rangeland. The most common interactions with cattle involve daily feeding, cleaning and milking. Many routine husbandry practices involve ear tagging, dehorning, loading, medical operations, vaccinations and hoof care, as well as training for agricultural shows and preparations. Interestingly, there are many cultural differences which occur when working with cattle; the cattle husbandry of Fulani men rests on behavioural techniques, whereas in Europe, cattle are controlled primarily by physical means, such as fences. In terms of food intake by humans, consumption of cattle is less efficient than of grain or vegetables with regard to land use, and hence cattle grazing consumes a larger area than such other agricultural production, especially when they are raised on grains. However, cattle and other forms of domesticated animals can sometimes help to use plant resources in areas not easily amenable to other forms of agriculture.

Cattle today are the basis of a multi-billion dollar industry worldwide. The international trade in beef for 2000 was over $30 billion and represented only 23% of world beef production. The production of milk, which is also made into cheese, butter, yogurt, and other dairy products, is comparable in economic size to beef production, and provides an important part of the food supply for many of the world's people. There are some pressing trepidations concerning cattle farming though. A report from the 'Food and Agriculture Organization' (FAO) states that the livestock sector is 'responsible for 18% of greenhouse gas emissions', and the report concludes, that unless changes are made, the damage may more than double by 2050, as demand for meat increases. Another concern is manure, which if not well-managed, can lead to adverse environmental consequences. These are issues which both farmers and governments are working on though, so that cattle farming and the commercial usage of cows can, from its long and largely inimitable roots, progress and develop into the future.

ON THE ECONOMY AND MANAGEMENT OF THE DAIRY.

CHAPTER I.

OF MILCH KINE.

THE value of the respective breeds of milch kine having been already stated,[1] it will rest with the farmer to make his selection, according to the nature of the soil, and the particular branch of dairying which he means to pursue. If his object is to sell milk, quantity must be the leading consideration; and quality, if he means to produce butter and cheese. Quality must not, however, be wholly sacrificed to quantity, in breeding cows for the milk-trade, for the law demands that milk supplied to the public shall have a minimum standard quality indicated by $8\frac{1}{2}$ per cent. of solids not fat, and $2\frac{1}{2}$ per cent. of fat,—together, 11 per cent. of solids. There are no breeds of cows in the British Islands whose milk, under ordinary conditions, will not yield upon analysis more than 11 per cent. of solids, even in early summer, when the grass is young, soft, and very succulent; though perhaps individual cows might be found whose milk would fall below that standard.

It is now admitted that the Channel Islands cattle—the Jerseys and Guernseys—yield richer milk that any other breed of cattle in the country, and this is equivalent to saying that they yield richer milk than any other breed in the world. Probably the little black Kerry cows of Ireland will be found to come next to them in quality of milk, closely followed by the blood-red cows of Devonshire and by the docile and picturesque Red Polls of Norfolk and Suffolk. Of Scottish breeds, the Aberdeen-Angus, the Galloway, and the Highland cows stand before the parti-coloured, hardy, energetic little Ayrshires, but the last-named more than compensate in quantity what their milk lacks in quality. The idea once persistently entertained by many people, that red cattle yielded the richest milk, may now be regarded

---

[1] See Book i. chap. i.

as archaic and altogether illusory. The breed of a cow, and her own individual qualities, have, along with suitable food, everything to do alike with the quantity and quality of her milk, whereas her colour of hair has little or no bearing upon the question.

There is a considerable difference in the milk-yielding capacity of different cows of any given breed, in respect to quantity and quality, one or both. It usually happens that milk is lowest in quality where it is highest in quantity, and *vice versâ ;* but no unvarying rule will be found to exist on these points. The "stream of tendency," however, is opposed to any idea which may be held that quantity and quality, in their highest development, are found in partnership, as a rule, in any cow or breed of cows. Nor does it by any means follow that the largest breeds of cattle yield the largest quantity of milk, and certainly it does not seem that they yield the largest total amount of solids. Individual cows of the larger breeds may perhaps be found whose yield of milk, and of total solids in that milk, are greater than those afforded by animals of the smaller breeds; but such cows are somewhat rare, and do not occur so frequently as to stamp the breed to which they belong with a dairy reputation corresponding to the size of the animals. And, indeed, it may be said that the richest milk is not always that which is yielded in small quantities, either by large or small cows, though at the same time it commonly is.

As a matter of fact, there is an almost infinite variability in the flow of milk in different animals of most breeds of cows, and it will probably be found that this variability occurs most generally among the larger breeds, though at the same time the smaller breeds are not by any means exempt from it. In every breed, there may be found tribes and families of cows which are much more famous for milk than is the breed at large; and any family or tribe bearing such a reputation owe it, as a rule, to the careful breeding and training, in that direction, to which they have been subjected. A large flow of milk is seldom maintained for a long period, yet, in reference to this, cows do vary very considerably from each other; for while some cows will hardly milk through half the year, it is with difficulty that others can be let dry for the next time of calving.

It will be seen, therefore, that scope enough exists for the care and energy of any one who, by careful selection and weeding out, has the mind to build up a herd of cows famous for quality and quantity of milk, in which these features shall become hereditary. It is, too, a work of time, of untiring attention, of sound judgment, and it can only be accomplished by a man whose heart is in his business. Breed and breeding are more potent than locality or country, in reference to the evolution of the milk-yielding function; and food and treatment have their influence as well,—an influence not inferior, perhaps, in the long run, to that of natural propensity. In building up a herd of cows great at the milk-pail, it is imperative that only such sires should be used as are known to come from cows and families that are excellent for milk, for the bull indeed is half the herd.

The test as to quantity of milk is easily made by keeping a record of the weight of milk given by each cow, each time she is milked. Details of the method are given in the description of the milk register on page 245. A spring balance (see fig. 51, p. 246), and a slate at the cow-shed, and Barham's "Sandringham Dairy Record Sheet," will enable anyone to obtain and preserve data which are of the greatest value—nay, indeed, are indispensable—where a herd is being improved for milk. In summer, when cattle roam over the pastures, where the food is the same for all the cows alike, each cow's capacity for milk may be exactly ascertained; and, in winter, the test may be made more searching still by noting the quantity of food consumed, as well as the quantity of milk yielded, by each cow respectively.

The test as to quality of milk, in reference to cream, is not so easy and simple, but it may be taken with a sufficient approach to accuracy by means of a set of graduated glass tubes, called "cream-gauges," which show the cream volume of as many samples of milk as may

Fig. 50.—Dairy Supply Co.'s Gerber Butyrometer.

be placed within them, in this way instituting a comparison between the milks of different cows. There are also small instruments, adaptations of the well-known cream-separators, in which several samples of milk may be tested for cream, in a few minutes' time. These ingenious machines are rapid in work, accurate in the results they obtain, and very easily turned by hand; they are, consequently, very well adapted to the needs of cheese- and butter-factories, or of any other institutions where it is desirable to test the quality of different milks received.

It may be contended, however, that a cream-test, for volume of cream, is not sufficient to denote the quality of milk, for cream varies in quality. The "Butyrometer" has been designed to ascertain the actual percentage of butter-fat in samples of milk, in order that milk may be bought on a basis of quality at butter-factories and other large

R

establishments. A number of samples may be tested at once in the machine, an illustration of which is given in fig. 50. The hand crank is gradually brought to a speed of fifty revolutions per minute, and this is maintained for three or four minutes; then the crank is left free, and the disk gradually brought to rest by gentle pressure with a cloth, after which the tests may be read off on the tubes. A hundred samples of milk may be tested in less than an hour, and the result is almost independent of the individual skill of the operator.

Machines of this kind are used, not only by wholesale purchasers of milk for sale as milk, but also generally in creameries, in which milk is purchased from dairy farmers to convert into butter. In Australia and New Zealand the price paid for milk in many creameries is in proportion to the percentage of butter-fat in each lot, and there is no doubt that this is the most equitable method of purchase.

Copious and long-continued lactation, wherever it occurs, is a natural function for the most part artificially developed. It is, in fact, the result of domestication of cattle, in the first place, by breeding and training, but to some extent by soil and climate; for it is found that cattle in a feral state do not yield more milk, or yield it for a longer period, than is necessary to give their offspring a good start in life. The quality of milk is largely a question of feeding, treatment, and climate, yet breed has more to do with it than all of these. We see this demonstrated in the Jerseys and Guernseys more than in any other breed of cattle; the superior quality of their milk is hereditary, and this heredity is no doubt owing to the care and tenderness with which these cattle have been treated for centuries, and to the genial climate of their island homes. The Jerseys, more particularly, exhibit the results of the influences mentioned, as will be seen from the table of figures in the next paragraph; and as these beautiful cattle have been most carefully bred, fed, and tended for a long period, we may accept the results as being conclusive in favour of careful breeding, kind treatment, and a genial climate. Whether or not this quality of milking in the Jerseys will be perpetuated in the breed, in other countries, and through succeeding generations, is a problem which only time can solve, but so far it betrays no sign of falling off in the United States, Canada, and elsewhere. All will depend on breeding and treatment, no doubt, for in any case the quality is now hereditary in the breed, and cannot be sacrificed save by unfavourable conditions.

For some years past, at their annual Dairy Show in London, the British Dairy Farmers' Association have conducted milking trials, and the results have been very instructive. These trials have been made in respect of quality as well as quantity of milk, and the results are summarised by Mr. P. McConnell, in Part I., Vol. VI., of the Journal of the Association, embracing a number of cows of different breeds and extending over a period of ten years. The trials were made by taking for analysis a sample of each cow's milk, at each morning's and evening's milking on one out of two days of the show. From the chemical data thus obtained calculations were made as to the butter-yielding capacity of each cow, and as to the butter ratio of the milk. The high quality of the milk of certain Jersey cows is remarkable:—One yielded milk containing 19¼ per cent. of total solids, of which 9½ per cent. was fat; the butter ratio being 11·4 lb. of milk to 1 lb. of butter. Another yielded milk containing over 17 per cent. of total solids, of which 8½ per cent. was fat; here the butter ratio was 11·3 lb. of milk to 1 lb. of butter, or at the rate of over 23 lb. of butter per week. The following tables shew the milking capacity of Shorthorns, Jerseys, and Guernseys,—the only breeds which have been tested by the Association in numbers sufficient to give a reliable picture of their value as milk-producers. But the few Devons, Ayrshires, Red Polls, and Kerries, which have been under test, give very satisfactory and promising results, as also does the single Welsh cow that has been entered:—

Breed.	Lb. of Milk per day.	Percentage of Total Solids.	Percentage of Fats.
119 Shorthorns	43·13	12·87	3·73
31 ,,	44·80	12·89	3·81
115 Jerseys	27·87	14·36	4·56
43 ,,	28·41	14·94	5·47
49 Guernseys	28·30	14·00	4·77
14 ,,	31·15	14·46	5·03

In connection with these figures it must be borne in mind that while the Shorthorns yield more milk, though not necessarily a larger aggregate of solids in the milk, than either of the other two breeds, they are larger cattle, requiring more sustenance, and a greater breadth of land per cow; they are, however, much more valuable as butchers' beasts when fat, and therefore more profitable when they go barren, or have seen the best of their days as milkers. Eleven Ayrshires, tested at the Dairy Show, yielded an average of 34·26 lb. of milk each per day, containing 13·43 per cent. of total solids, of which 4·15 was fat; as these are small cattle, hardy, energetic, suitable for inferior land, and for trying climates, it will be seen that they are among the most valuable of our milking breeds.

Fed and treated similarly, it will generally be found that cows yielding the smallest quantity will afford the richest quality of milk; but this is not by any means an unvarying rule, and we sometimes meet with very striking instances to the contrary. Both quantity and quality frequently vary, in any cow of any breed, or as between any two or more cows of any given breed, of the same age, of similar size and constitution, fed on the same quality and quantity of food, and so on. The state of health of the cow, changes in the weather, in food, in treatment, the period of the year, the time which has elapsed since calving, the degree of succulency and digestibility of the food, the gentleness and attention which are bestowed, &c., &c., have each and all a distinct though perhaps not sufficiently appreciated influence on the flow and quality of the milk. Good old pasture land, not necessarily the richest, but sound land, with a good assortment of indigenous grasses, improved if needful by judicious top-dressing, will as a rule yield the best qualities of both cheese and butter; but rank pastures, sewage grass, succulent green crops, and so on, are not well adapted for either purpose. Turnips, mangel, brewers' grains, cabbage, &c., will increase the quantity of milk; but its quality is best improved by leguminous meals, ground oats and maize, which indeed may be fed to milking cows with advantage along with grass through the summer and autumn.

It has been thought that food rich in oil and carbo-hydrates would yield the richest milk; this is not the case, however, for while such carbonaceous food will increase the quantity, albuminoids or nitrogenous food will best improve the quality. The carbonaceous food is well represented in linseed, potatoes, and mangel, which are rich in oil, starch, and sugar respectively; and the nitrogenous food by broken beans and peas, vetches, clover, and their allies.

All the same, however, it is the quantity rather than the quality of milk which the sooner responds to better and increased food, though its quality too will improve when the limit of expansion as to quantity has been reached in this way. Regular feeding on good food will yield more satisfactory results than that which is spasmodic and irregular. The casein in milk varies less than the fats in amount, and while food rich in carbonaceous ingredients is more likely to influence the quantity of butter-fat in milk than nitrogenous food is to alter the proportion of casein, either kind of food will most of all increase the quantity of milk and the proportion of butter in it. Lean cows will yield less and poorer milk than those which, without being actually fat, are kept in good store condition, and the milk of all cows begins to decline in quantity, and to improve in quality, after the first three months of lactation.

Careful investigations into the effect of changes of food on the yield of milk have been repeatedly made, and the following conclusions may be regarded as broadly and fairly established:—Firstly, an increase of food, sustained in both quality and quantity, increases the yield of milk, and also the proportion of solids in it, and the better milker a cow naturally is the greater will be the effect of the foods. Secondly, the proportion of fat in the food bears no special relation to the proportion of fat in the milk, but an increase of fat in the food increases the yield of milk as a whole. Thirdly, while albuminoids from their nature have been supposed to be specially adapted to increase the proportion of casein in the milk, it has been found that a liberal use of them tends more to an increase in the proportion of fat, for casein varies very little as compared with fat in milk. Fourthly, the composition of milk as regards any one of its ingredients does not respond, with anything like fidelity, to changes made in corresponding ingredients in the food the cow eats, and scientific feeding is followed by such uncertain results, save with respect to the increase of the total yield of milk, and, consequently, of the total solids in it, as to preclude the laying down of any definite rule concerning it. The composition of milk, in fact, primarily depends more on the breed, or on the capability, of a cow than on the food she eats; and the limit of milk production is soon reached in a cow not naturally given to much milk, however rich the food may be, whereas the effect of a plentiful supply of good food on a cow naturally inclined to milk is, as a rule, very considerable.

The need of greater exactitude in the dairy has led to the introduction, within recent years, of the Milk Register. By a milk register is simply meant a record of the quantity of milk yielded by a cow. In other words, it is a quantitative estimation of the milk the cow gives. It affords no information as to the quality of the milk, or as to its butter-yielding or cheese-yielding capacity. Nevertheless, by its means, the milk-producing capacity of a cow can be ascertained exactly, and her character in this respect can be expressed by means of figures about which there can be, or should be, no equivocation. A greater or less degree of exactness can be secured, according to the greater or less frequency with which the register is taken. A register based upon observations made only once or twice a week would be less instructive, and, in a sense, less valuable than a register based upon observations made once or twice a day, though it is by no means implied that even a weekly register would not prove extremely useful.

In the taking of the register two methods suggest themselves, and the question arises, which is the better, that by volume or that by weight? Against the volumetric estimation in, say, gallons, quarts, or pints, there are several objections, amongst which are the trouble of pouring the milk into the measures and the difficulty of allowing for the froth. Chemists, who have brought the art of estimating quantities to great perfection, invariably employ gravimetric methods, even their so-called volumetric processes being based ultimately on proportions by weight. Hence, it seems desirable to estimate the quantity of milk by weight rather than by measure. Moreover, the transition from weight in pounds to the equivalent measure in gallons is easily effected in the case of milk. Thus, the specific gravity of milk being 1·03, it follows that 103 lb. of milk will occupy the same space as 100 lb. of water, but for practical purposes these numbers may be taken as identical. Then, since one gallon of water weighs 10 lb., no appreciable error is involved in considering that one gallon of milk also weighs 10 lb. Consequently, if the quantity of milk given by a cow at one milking be expressed in lb., it is only necessary to place the decimal point on the left of the unit figure to get the equivalent in gallons. Thus 24 lb. of milk represent 2·4 gallons, 15 lb. of milk represent 1·5 or 1½ gallons, and so on.

The practice of taking the milk register, as followed in a dairy well known to us, may be described. The cows are always milked in the stalls, and during summer they are brought in twice a day for this purpose. After each cow is milked, the pail containing the whole of her milk is hung on a spring balance suspended in a convenient position, and from the gross weight indicated there is deducted the already known weight of the pail. The difference, which represents the weight of milk, is recorded in a book suitably ruled. This book when open presents a view of one week's records. In the left hand column are the names of the cows; on the right of this are fourteen columns, two of which receive the morning and evening record of each cow. In a final column on the right appears the week's total yield for each cow, and space is also allowed for any remarks. Fractions of a pound are not entered, but 18 lb. 12 oz. would be recorded as 19 lb., whereas 21 lb. 5 oz. would appear as 21 lb., so that a fraction of over half a pound is considered as a whole pound, and a fraction of under half a pound is ignored.

The need of deducting the weight of the vessel is obviated in such a simple appliance as that of the Dairy Supply Company, illustrated herewith (fig. 51). It is easily movable from place to place, and shows

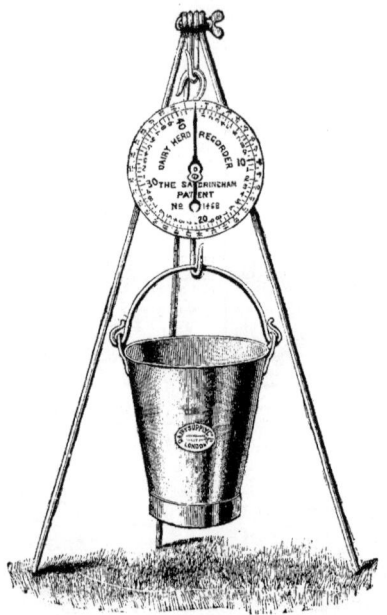

Fig. 51.—Milk weighing Appliance.

on the dial the weight in pounds and ounces, as also the measurement in gallons and pints. As the pail weighs with the machine, no deduction for it is necessary.

Every dairy farmer has some idea, as to each of his cows, whether she is a good, a bad, or an indifferent milker, but such knowledge is at best only vague. By the simple means indicated, the character of each cow as a milk-producer is slowly but surely recorded in a manner which is at once exact and definite.

As such a record affords no information regarding the quality of the

8

milk, it is obviously of most use to dairy farmers engaged in the urban milk trade. It is, moreover, particularly valuable to the farmer in that it shows to him the relative milk-yielding capacities of his cows, and thus enables him to gradually weed out the naturally poor milkers, and replace them by better ones.

The study of a milk register extending over, say, a year is most instructive. The influence of external conditions on the discharge of the lacteal fluid is clearly indicated. A change in the food supplied to the cows, for example, or the appearance of sharp frosts, as well as other sudden meteorological changes, will be found to be faithfully reflected in the milk-pail. The register may bring out, amongst other circumstances, some such useful fact as the following : that of two cows, for instance, one may be notorious for giving at times large flows of milk never approached by the other cow, and may have acquired the reputation of being the better milker, while the register may prove that, when duration of lactation and average yield of milk are considered, the second cow is decidedly the superior. The register will, in fact, indicate unerringly which are the best milk-yielding cows in the dairy, and which therefore are, with this object in view, the best to breed from. If it is desired to know the richness in cream, then the lactocrite, or some simpler instrument of which there are several kinds, may be called into requisition.

The simplicity and inexpensiveness of the milk register must not be overlooked. These are features which should commend it especially to the notice of small dairy farmers, for with a moderate number of cows it is particularly easy to introduce the register. But even with a large dairy it will be found that, as soon as the system has got fairly established, the additional time and trouble involved will sink into insignificance when compared with the benefits which cannot but accrue from the intelligent study of a faithfully kept register.

CHAPTER II.

Of the Pasture and other Food best suited to Milch Cows.

THE feeding of milch kine comprises two distinct methods, viz., *pasturing* and *house-feeding*.

In order to obtain an abundant supply of good milk, where the pasturing of cows is adopted, it is not alone requisite that the grass shall be plentifully produced, but also that it shall be of such quality as will be relished by the cattle; and this property will generally be found in old natural pastures that have been properly managed. Excellent pasturage is, however, provided on the lighter soils, by new "seeds" which are taken in rotation, and kept down for several years. In the important dairying counties of the south-west of Scotland, large herds of Ayrshire cows are pastured to a great extent on such newly-seeded land, rye-grass being the chief component, and the cheese and butter made from the milk are of excellent quality. New pastures are not always considered to produce the best milk, but on light soils they are commonly superior to old turf; and while in some old pastures there is too generally a large proportion of weeds—including various composite and umbellate plants—which impart a flavour to the milk, the newly-seeded land is usually free from them, because all such plants have been destroyed by cultivation. The newly-seeded land, too, if it has been properly cleaned, is free from the following, which are sometimes found in old pastures:—Hair-grass, Yorkshire fog, quaking-grass, brome-grass, buttercups, plantains, and other plants. These are weeds, cattle do not like them, and they should be eradicated by lime and other top-dressings which sweeten the soil and promote the growth of the better grasses.

Some old pastures, indeed, are so foul with weeds and inferior grasses that to plough them up, take a crop or two of corn, and a summer fallow, or a crop of roots for which the land has been well cultivated, would seem to be the quickest and surest way of exterminating the intruders. On the dry, open soils, this may be done with advantage, perhaps; but on the heavy retentive soils such a course is obviously out of the question on the ground of expense. Top-dressing, indeed, with lime, or with superphosphate of lime and kainit, affords the best solution of the heavy land problem, so far as permanent pasture is concerned. It is likewise worthy of note, that although the long rank grass, growing in orchards or other places, in general feeds well, and produces a flush of milk, yet such milk will neither be so rich, nor carry so much cream in proportion, as the milk of the cows that are fed upon short fine grass; nor will the butter be so good.

The quantity and quality of milk are materially affected by driving cows a long distance from one pasture to another; hence it will be proper to have the steading in as central a part of the farm as possible. It is also of essential importance to have pastures well sheltered and enclosed, as the produce of milch kine will be greatly improved, or deteriorated, according to the attention or disregard bestowed on this point; for, when confined within proper inclosures, they not only feed more leisurely, and are better protected against bad weather, but are also less liable to disturbance than when they wander into other fields. Shelter from the rays of the sun; immunity from being chased about by flies, or dogs, or foolish people; and moderate warmth and quiet are also greatly conducive to an increase of milk.

With regard to the housing of milch kine during summer, a common practice is, where there are proper enclosures, to send them out in the evening, in order that they may lie out during the night, while in the heat of day they are kept more cool and quiet in the cow-sheds than in the fields. The advantages resulting from this course are obvious, for the cattle obtain in the night the exercise which is beneficial, and in the day they are not scorched by the rays of a hot summer's sun, nor are they tormented by the flies, especially the ox warble flies, that are so active in the daylight.

The most general practice in the British Isles, particularly in the Midlands and the North, is to have cows out in the pastures, day and night alike, from May to November. Tethering is not employed very much anywhere, save in the Channel Islands and in a few places in the south of England; it involves a good deal of trouble, for the cows need watching, and moving, and watering, pretty frequently, in addition to milking. The system economises grass, no doubt, but not to so great an extent as in the system called "Soiling,"—that is, cutting all the grass and green crops, and carting them to the sheds, to be consumed there by the cows. There can be no dispute on the point of waste of grass, where cattle roam at large on the land, though the actual waste is less than many men think, except in wet seasons. It becomes, indeed, a question of relative profitableness, as to whether the waste in grazing is equal to the cost of cutting and carting the food to the sheds. This is a problem which every farmer may solve for himself, according to circumstances. But in any case there can be no question that cattle are healthier on the pastures than in the sheds; and as they must be in the sheds during the winter, it is perhaps best, all points considered, that they should be on the pastures when they may, in spring, summer, and autumn.

In boisterous weather, whenever it may occur, cows should be sheltered, or else they will at once fall off in milk. But the shelter need not necessarily take the form of sheds, if only good fences and plantations are available. In winter there is no alternative, and cows must be housed all the time, save in warm and sheltered localities. Exercise, however, even in winter, is always a good thing, if only in walking a short distance to the water, twice a day. Some people advocate open sheds for dairy cattle, attached to warm yards; others loose boxes, one for each cow; and yet others recommend stalls, in which the cows stand side by side, tied by the neck. The latter plan economises litter and room much better, and on the whole is cleaner, than either of the others. Cows in stalls need no litter to lie on, even when the stalls are paved with stone or brick,—if only a layer of clay be put under the fore-feet, to soften the place for her knees, when the cow lies down and rises up. Litter, indeed, in the form of straw, is too valuable, as a rule, in these days, to be used for cattle to lie upon, and it may be declared with authority that they will do very well without it.

Cow-houses are variously arranged. The most convenient are known as double sheds, under which arrangement two rows of cattle stand tail to tail, and each of these rows head to head with another row; where they are tail to tail, a roadway and two manure gutters are between them, and where head to head a gangway or "fodder bing." The dimensions of the stalls may be the following, for large cows: seven feet long, including manger, and six feet six inches wide; this will serve for two cows, with a short partition between them at the manger. The space occupied by the two manure gutters and the path between them should be seven feet wide. For medium-sized cows, the stalls may be reduced half a foot each way; and for small ones, more in proportion. The "fodder bing" may be any width that is desired—big enough even for a hay barn, if need be; or it and the cow-sheds too may be floored overhead, forming hay-lofts for storing forage. This latter arrangement is, however, now considered more or less objectionable, and other means of storing fodder should, if possible, be provided. In the plan (fig. 52), it will be seen that provision is made for eighteen cows, twelve in a double, and six in a single shed, all under one roof; and it will be obvious that the single shed might be made a double one by simple lengthening of the building. On the whole this arrangement is the best of all, for the doors all open into the yard, and the manure sheds are all out of it, while every convenience for feeding the cattle is provided. It is perhaps as well to say that in order to preserve the volatile and more valuable ingredients of the manure, roofed sheds

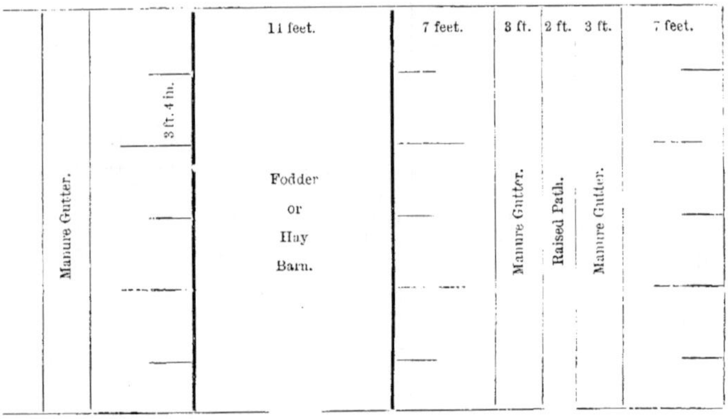

	11 feet.	7 feet.	3 ft.	2 ft.	3 ft.	7 feet.
Manure Gutter.	3 ft. 4 in.	Fodder or Hay Barn.		Manure Gutter.	Raised Path.	Manure Gutter.	

Fig. 52.—Plan of Cow-house.

should be built to contain it until it is taken out on the land. This is the system on which, after long experience and thought, we put up a new set of farm buildings, and managed thereby to secure warmth, light, ventilation, cleanliness, and convenience—the chief desiderata in providing accommodation for cattle.

In the management of milch cows, it is important that they should be kept in good health and in fair store condition all through the year, and particularly in winter—the period when they depend on house-feeding, and have no chance of helping themselves to what they can find. If they fall away in flesh during the winter, from insufficient food, food of inferior quality, exposure to cold and damp, or other causes, they cannot yield as much milk in the ensuing summer as they will if they were in good condition, and it takes them a good part of the summer to "get their backs up again," as the saying goes. Cows should not be lean when calving-time comes on; for, in case they are, the feeding they may get will have to be very liberal indeed to bring them up to a full yield of milk.

Farmers, as a rule, do not like feeding cows liberally when they are dry for calving, and are therefore yielding no return; yet that is just the time when they can best be got into condition again, ready for another period of lactation. The practice too commonly is to feed them on straw, rough hay, and other inferior stuff that is rightly considered not good enough for them when they are yielding milk; but then, such food is not good enough for them at any time, unless it be improved by the addition of corn, and spice for flavour, or cake. All inferior forage should be improved in some way, and the time to use it up is certainly not when cows are dry for calving, unless it be improved by corn or supplemented by cake; the common practice of keeping dry cows on inferior forage during the period indicated is a mistake, the consequences of which are seen later on, and it seems strange that people can be found who still persist in it. During the winter, therefore, in-calf and in-milk cows should receive food that is nutritious, as easily digestible as possible, sound and good of its kind. Straw alone will not do, and even good hay may with advantage be supplemented by a couple of pounds of cake per day; it is surprising what a difference even this small quantity will make, used week after week and month after month, through the winter, especially when the cows are dry. When cows are fed on straw or coarse hay, alone, without any food more generous in character, the organs of lactation become more or less attenuated and inert, and are not easily or quickly restored to what under generous feeding would be their normal condition. With inferior forage, at least 4 lb. of cake should be given.

It is essential that milch cows should always be kept not only in good store condition, but in a generous state of milkiness, ready and free to work, like a well-oiled machine. A few swede turnips, or some mangel, each day will be found very useful to this end, and two pounds of linseed cake, or boiled linseed on chaff, increased to four pounds when the cow has calved, will keep her in suitable condition, add to the strength of her constitution, stimulate the digestive organs, and enable her to make the best use of the succulent grass of early spring.

In Holland, where the management of cows is carried to the highest perfection, the animals are curried in the same manner, and kept as cleanly, as horses in a stable.[1] If this is an error, it is at least one on the right side, and the invariably high condition of all Dutch dairy stock is the surest proof of their superior management, the chief features of which are,—care in keeping the cattle dry as well as clean, food suitable and adequate to their requirements, and attention to the purity of the water. This last-named point is considered of such importance that the water is not even suffered to be tainted by the breath of the beasts. And yet it is a known fact that cattle frequently prefer the water of ponds impregnated with the urine of other animals;

---

[1] Baron d'Alton, in "Communication to the Board of Agriculture," vol. i.

a circumstance probably arising from the saline matter which this water contains, and which instinct points out as beneficial to their health.

The extraordinary cleanliness and neatness, which have in the course of centuries become hereditary habits in the Dutch people, are found to prevail everywhere, and in everything, in Holland—in the fields, fences, roads, plantations, houses, buildings, as well as in the management of cattle. On several occasions we have noticed this with much interest, and with regret that such customs do not similarly prevail in other countries. The grooming of milch cows, however, and their better treatment in many ways, is spreading to other lands, and in England, Scotland, Germany, Canada, the United States, and even in Mexico, as we can also testify, dairy cows are treated with great kindliness,—well fed, well groomed, well cared-for generally,—in instances numerous enough, perhaps, to act as a leaven that "will leaven the whole lump." In the south-western counties of Scotland, where the plucky little Ayrshire cows are mostly found, the art of cattle management, and of dairying in all its features, has attained a high degree of excellence; a week or two spent among the dairy farmers in the neighbourhood of the Mull of Galloway, will reveal a condition of things, as indicated, not easily equalled in many other portions of the British Islands.

It has already been intimated that the best summer food for cows is good grass, spontaneously growing on sound land; but when such grass is limited, or failing, then tares, lucerne, and clover, either cut or pastured, may be very advantageously used as supplementary food. There is a prejudice against tares, from their being supposed to render the milk *ropy;* but we have been assured by a farmer who kept twenty-one cows of a mixed breed on the verge of Epping Forest, that he soiled them night and morning on tares during a great part of the summer, without any other assistance than the common pasture of the forest, and that not only was there no appearance of ropiness in the milk, but it was far richer than when the cows were fed on meadow grass, the butter likewise being of the finest quality.

Beans given in conjunction with good pasturage are excellent for keeping cows in milking order, and also in good condition. The beans should be kibbled, and from three to four pounds of the broken material given per day.

Good sweet hay is the staple winter food of a milch cow; the accessories are those usually employed in feeding and fattening cattle. Swede turnips, beans or peas broken, and oil-cake, will render the milk richest. But carrots, mangel, and potatoes may be given.[1] Indeed, on the Continent the mangel is preferred to other roots for feeding cattle,[2]

[1] In the Island of Jersey, about 35 pounds of parsnips are given daily to the cows, with hay. They are found to improve the quality of the cream, which is more abundant than from an equal quantity of milk from cows differently fed—*seven quarts producing as much as seventeen ounces of butter*—and the flavour of the latter is superior.—"Quayle's General View of the Norman Isles."

[2] Mr. Harley, at his dairy at Willowbank, put the comparative value of mangel and Swedish turnips to the test. He took an equivalent weight of each, and gave them to

and many accounts are given of the nutritive powers of the potato ; one bushel, *per diem*, with good meadow hay, is said to cause a milch cow to yield as much milk as she would when fed on the finest pasturage.

Turnips of all kinds are apt to flavour the milk more or less unpleasantly ; this, however, it is said, may be prevented by cutting off the crown of the turnip, and giving only the lower portion to the cows ; by pulping the roots, steaming or cooking them ; and by always giving them to the cows immediately after milking. The flavour of turnips is volatile, and may therefore be got rid of under these precautions. It is claimed that placing a small piece of saltpetre in the milking pail will counteract the odour of turnips.

Cabbages are of great service, but they require to be given with a considerable portion of sweet hay ; and, like turnips, are apt to impart an unpleasant flavour to butter, unless great care is taken to remove all the decayed leaves. Kohl-rabi appears to be less objectionable in this respect. Fog, or rowen grass, is reserved for use in late autumn and winter. To these may be added, as generally useful in winter, pulverised oil-cake, linseed jelly, and grains,[1] all of the latter, as indeed any kind of meal, to be used with chaff which has been covered up, moistened with scalding water, and left for several hours to cook. By a judicious use of these various articles, together with a due mixture of dry food, considerable nutriment will be thrown into the system, the regular secretions will be excited, and the quality of the milk very materially improved.

But in some districts, farmers object to the use of roots or green food for milch cows, alleging that it spoils the milk ; they feed this portion of their stock entirely on the best hay and oil-cake during the winter.

Malt has been highly recommended, the animals fed on it being said to yield better flavoured and richer milk than can be obtained from cows kept on roots or cabbages. The expense, however, will always prevent this article of food being used to any great extent.

Steamed food is generally admitted to produce more and better milk than raw. This can hardly be due to increased digestibility ; but where hay is mouldy, the fungoid growth can only by steaming be rendered harmless ; and, indeed, such hay should always be steamed whenever it may be found necessary to use it at all as food. But with regard to hay and straw generally, steaming is not at all necessary to

---

two lots of cows of equal numbers, great attention being paid to the quantity and quality of the milk produced, and the improvement in the condition of the cattle. In these respects, however, there was found to be little or no variation. The quantity and quality of the milk and the improvement of the cattle were much the same ; but the Swedish turnips were ultimately preferred on account of the deep soil which the mangel required.--"Harley's Dairy System," p. 71.

[1] Mr. Harley thus speaks of *grains:*—"When they were plentiful and cheap—which was generally the case in winter—a large portion of them were given with the more succulent food, but they were apt to make the cattle grain-sick. It has been ascertained that, if cows are kept long upon grains or distillers' wash, their constitution will soon be destroyed, and cattle thus fed should not be kept longer than eight or ten months. A little boiled linseed was considered to be the best antidote in preventing distillers' wash from injuring the health of the animals ; and wheat-straw, cut short and mixed with the grains, prevented the cows from being grain-sick."—"Harley's Dairy System," p. 74.

increase the digestibility by softening the fibre; the same end may be attained much more easily and cheaply by simply moistening the forage well with water, and leaving it a day to soften.

In some parts of Flanders, after the corn crops have been reaped, the ground is lightly ploughed and sown with spurrey. The cows are tethered on it in October, and a space allowed to each proportioned to the crop and the size and appetite of the animal. The butter from the milk thus obtained is called *spergule butter*. It is not of equal quality with that produced from the common food.[1]

In the midland and northern counties, milch cows are allowed the best pastures during summer, followed in the autumn by eddish, and various green crops, of which cabbage is regarded as one of the most important; and are housed for the nights when the weather becomes cold or wet, one or both, when they receive the first instalments of winter food, which in former days was hay, or turnips and straw where both were cultivated: but a difference was made between those which were rather fresh of milk, and those which were nearly dry, the former having a larger portion of turnips, with the addition of hay, while the latter were put off with little else than chopped straw until within a few weeks of calving, when hay was allowed. In Essex, the system was nearly the same, except that, the produce of the dairy being chiefly butter, turnips were seldom given. Rowen (or aftermath) hay, as being the softest and greenest, was preferred, and the consumption was calculated at two loads (of eighteen cwt.) in the winter, with two acres of summer pasture, and some straw, while drying off.

In the neighbourhood of London, distillers' grains and wash are extensively given to milch cows, and with advantage as regards the quantity of the milk; these articles do not, however, improve the quality. Grains are very liable to fermentation, and fermenting food is injurious to cows.

The vast extension of the country milk-trade has done away with most of the metropolitan cow-sheds, and has changed the character of dairy farming in districts from which milk is sent to London and other large centres. There can be no doubt that the milk-trade has been on the whole more profitable than cheese- or butter-making since 1878; indeed, it may be regarded as being and having been the mainstay of the dairying industry of England. It has led to a vast consumption of purchased feeding-stuffs, and therefore to more generous and liberal rations for cattle, as well as to more elastic and adaptable systems of cropping arable land. Later on we shall have more to say on the subject of the country milk-trade.

In the course of the preceding statements, the *stall-* or *house-feeding*, of cows during the winter in Holland has been mentioned; and from the remarks of Baron d'Alton, it appears that this method of feeding is there adopted *throughout the year* with greater profit than can be obtained from pasturing. The Baron, certainly, says that cows must be early trained to the confinement of stall-feeding, otherwise they do

[1] Sir John Sinclair's "Hints on the Agriculture of the Netherlands, &c."

not thrive; but, as the advantages of soiling and stall-feeding are so great, there can be no difficulty in adopting it, and, where it is intended to keep cattle thus, the calves may be easily reconciled to the confinement from an early age.

Mr. Horsfall's system of dairy management, recorded with such fulness and accuracy of detail in the "Journal of the Royal Agricultural Society of England" (Vol. XVII., p. 260, First Series), has deservedly attracted considerable attention. We give here his own description of it:—

"My dairy is but 6 feet wide by 15 feet long, and 12 feet high; at one end (to the north) is a trellis window, at the other an inner door, which opens into the kitchen. There is another door near to this, which opens into the churning-room, having also a northern aspect; both doors are near the south end of the dairy. Along each side and the north end, two shelves of wood are fixed to the wall, the one 15 inches above the other; 2 feet higher is another shelf, somewhat narrower, but of like length, which is covered with charcoal, the deodorising properties of which are sufficiently established. The lower shelves being 2 feet 3 inches wide, the interval or passage between is only 1 foot 6 inches. On each tier of shelves is a shallow wooden cistern, lined with thin sheet lead, having a rim at the edges, 3 inches high. These cisterns incline downwards, slightly towards the window, and contain water to the depth of three inches. At the end nearest the kitchen, each tier of cisterns is supplied with two taps, one for cold water in summer, the other with hot water for winter use. At the end next the north window is a plug or hollow tube, with holes perforated at such an elevation as to take the water before it flows over the cistern.

"During the summer the door towards the kitchen is closed, and an additional door is fixed against it, with an interval between well packed with straw; a curtain of stout calico hangs before the trellis window, which is dipped in salt water and kept wet during the whole day, by cold water spurted over it from a gutta-percha tube. On the milk being brought in, it is emptied into bowls. (The bowls are of glazed brown earthenware, standing on a base of 6 or 8 inches, and expanding at the surface to nearly twice that width. Four to five quarts are contained in each bowl, the depth being 4 to 5 inches at the centre.) Some time after these bowls have been placed on the cistern, the cold water taps are turned till the water rises through the perforated tube, and flows through a waste pipe into the sewer. The taps are then closed so as to allow a slight trickling of water, which continues through the day. By this means I reduce the temperature, as compared with that outside the window, by 20°. I am thus enabled to allow the milk to stand till the cream has risen, and keep the skim-milk sweet.

"Having heard complaints during very hot weather of skim-milk, which had left my dairy perfectly sweet, being affected so as to curdle in cooking on being carried into the village, I caused covers of thick calico (the best of our fabrics for retaining moisture) to be made;

these are dipped in salt water and then drawn over the whole of the tin milk-cans; the contrivance is quite successful, and is in great favour with the consumers. I have not heard a single complaint since I adopted it.

"Finding my butter rather soft in hot weather, I uncovered a draw-well which I had not used since I introduced water-works for the supply of the village and my own premises. On lowering a thermometer down the well to a depth of 28 feet, I found it indicated a temperature of 43°,—that on the surface being 70°. I first let down the butter, which was somewhat improved, but afterwards the cream; for this purpose I procured a movable windlass, with a rope of the required length; the cream jar is placed in a basket 2 feet 4 inches deep, suspended on the rope, and let down the evening previous to churning. It is drawn up early the next morning, and immediately churned; by this means the churning occupies about the same time as in winter, and the butter is of like consistency. The advantage I derive from this is such that, rather than be without it, I should prefer sinking a well for the purpose of reaching a like temperature.

"When the winter approaches, the open trellis window to the north is closed, an additional shutter being fixed outside, and the interval between this and an inner shutter closely packed with straw, to prevent the access of air and cold; the door to the kitchen is, at the same time, unclosed, to admit warmth. Before the milk is brought from the cow-house, the dairymaid washes the bowls well with hot water, the effect of which is to take off the chill, but not to warm them; the milk is brought in as milked, and is passed through a sile into the bowls, which are then placed on the cisterns. A thermometer, with its bulb immersed in the milk, denotes a temperature of about 90°. The hot water is applied immediately, at a temperature of 100°, or upwards, and continues to flow for about five minutes, when the supply is exhausted. The bowls being of thick earthenware—a slow conductor—this does not heighten the temperature of the milk. The cooling, however, is thereby retarded, as I find the milk, after standing four hours, maintains a temperature of 60°. This application of hot water is renewed at each milking to the new milk, but not repeated to the same after it has cooled."

At the Aylesbury Dairy Company's Farms, at Stammerham, Horsham, Sussex, the staple of the food given to cows in the milking-houses is ground oats, which produce very sweet milk. From 5 lb. to 6 lb. of crushed oats, with 1 lb. of other meal—wheat, pea, barley, or maize, according to season and convenience—are given to each cow per day. The meal is mixed with chaff, and with pulped or sliced mangel or cabbage, or with silage, according to season. Over this mixture, a kind of soup, made by boiling linseed, at the rate of 1 lb. per cow in winter, and half that quantity in summer, is poured, and the whole is allowed to heat slightly, the mixture being prepared a day before it is used. Salt, at the rate of 2 oz. per cow per diem, is sprinkled over the food. When silage is used, the quantity given is from 5 lb. to 8 lb. per cow each day. There is an ample supply of water, which is never allowed to stagnate in the troughs. Indeed, as food and water are given in the same troughs, there cannot be either refuse food or stagnant water, as the unconsumed food must be taken away before the water is turned on, and the latter, of course, must be allowed to run off before the food is put in.

A system of preserving grass and other green food in a succulent condition—ancient in Eastern Europe,—has within recent years been introduced into England. In Hungary the practice of storing forage, and even grain, in pits dug in the earth, has been followed from prehistoric times. Pliny says it was adopted in Greece and Spain, and even in Africa. In the "Journal of the Highland Society" in 1843, Professor Johnson gave a description of the German system of making "sour hay;" and in the "Journal of the Royal Agricultural Society of England," for 1874, it is described as seen in the East of Europe, where the green grass or green maize was crammed tightly down into long trenches four feet wide by six or eight feet deep, and covered over with a foot of earth. During some years little or no notice was taken of the subject in this country, and it was left for the wet season of 1879, in which hay-making was almost an impossibility, to cause English farmers to grasp at any alternative that was within their reach.

In the United States, claims of an extravagant character were made on behalf of ensilage, but as the result of trials made at the Missouri Experiment Station it was concluded that the air-drying method, with dry storage in a good barn in a compact form, is more economical than storing in the silo. Missouri farmers are not advised to build silos until there is a radical change in conditions.

The system, in point of fact, is an alternative to hay-making, and, notwithstanding the circumstance that silage is in a more easily digestible condition than hay, it is probable there are few farmers, to whom farming is something more than a hobby, who will persevere with making of silage if only they can make good hay. In giving evidence on the subject, Sir John Lawes said " his past experience had caused him to form the opinion that a ripe crop of oats, being cut up, straw and corn mixed, produced more meat than the silage. The chief value of silage consisted in its storing, thus producing food available at all times.

" Silage-fed milk was richer to look at and taste, but still they could not trace that the animal fed on silage had made so much butter as that fed on mangel. It was desirable in making silage to avoid chemical change in the silos as much as possible, because fermentation meant loss. In some of his oat silage, as far as they had cut it, the analysis showed a loss of nearly 30 per cent.

"Asked whether, after his experience, he would now go to the expense of building a silo if he had not got one, witness said he was rather doubtful about it. It was very useful, no doubt, but he was not quite sure that he would go to the expense. He could not do without roots. He had not grown buckwheat, and he had no opinion of it as a cleaning crop. Winter oats cleaned land better than any corn crops that he knew of. He had not been able to grow maize; but for

s

ensilage a maize crop was everything, because it cleaned the land at the same moment. He preferred clover silage infinitely to the oat silage. Sweet silage seems suddenly to go bad frequently; but he had never made any."

One of the most careful farmers of our acquaintance has dropped the system after several years' trial of it, and after forming a favourable opinion of grass silage as a supplementary food for milk cows. Silage, indeed, even when successfully made, can only be used as a subsidiary food for stock, in the place of roots to some extent. Hay, and swedes, or mangel, will be found preferable to hay and silage. To say that silage is better, more nutritive, than the grass from which it is made, is hardly compatible with common sense. In the silo there is fermentation,—sometimes a good deal of it,—and where fermentation occurs there is a loss of nutritive matter. Probably the succulency of silage, as compared with hay, is its chief merit, and it is no doubt useful where no roots are grown. Too much of it will injure the milk, and it may be expected that cows fed extensively on it for two or three years, will, as when fed continuously on brewers' grains, hardly be improved in constitution. It is the writer's opinion that if farmers can make good hay, and will moisten some of it—not soak it—for a few hours, before feeding it to the cows, they will feel no need of silage. All depends, in fact, on whether good hay can be made, and it must be borne in mind that ensilage is not always a success. The subject of ensilage is discussed more at length on page 842.

It is maintained by the best breeders that the mixture of salt with the food is beneficial to the health of stock. Some, indeed, have a lump of rock-salt placed in the manger, at which the cattle may lick when they feel inclined. This is a practice which we strongly recommend.

CHAPTER III.

Of the Situation and Buildings proper for a Dairy—Dairy Utensils.

A DAIRY-HOUSE ought, if possible, to be so arranged that its lattices may never front the south-east, south, south-west, or west. A northern aspect is the best, and there should be openings at each end of the building, in order to admit a free current of air. These lattices, which are in every respect superior to glazed lights, may be covered in summer with gauze wire, perforated sheet-zinc, or oiled paper pasted on pack-thread stretched for that purpose, so as to admit the light, whenever it may be necessary to exclude cold winds. A perfect milk-room is one that is dry, clean, cool, well-ventilated, free from atmospheric impurities, and uniform in temperature.

The *situation*, for the sake of convenience, should be near the cow-house as well as the farm-house; but care should at the same time be taken that it is not exposed to the effluvia of the cow-house, stables, or farm-yard, as any bad odour might taint the milk, and give an unpleasant flavour to the butter.

If it can be so managed, the dairy should be well sheltered by trees or by the situation of the ground, on the north, the south, and the east. The grand principle of its construction should be to preserve, as much as possible, an equal temperature both in summer and in winter. This is managed in Switzerland and in some parts of France by the dairy being constructed in the heart of a rock. In Ireland and elsewhere the same result is attained by having double walls and a double roof, with a free circulation of air. The second, or upper roof, may be of roof-felting on a light frame of wood; the object is to secure an "air-cushion," as a non-conductor of heat. In Switzerland the business of the dairy is removed as far up the mountain as convenience will permit, and sometimes, at a considerable distance from the cow-house and the residence of the farmer. A pump, or other source of pure water, should open into the dairy. In a level country, however, like those districts of England in which our largest dairies are found, it will, as above stated, be for the convenience of the farmer to have the dairy as near to the cow-house and his own residence as possible, but while there may be proximity, there should be no direct communication between the cow-house and the dairy.

Where the produce of the dairy is the principal object in view, a little extra expense in the construction of the dairy-house will be ultimately more than repaid by the superior quality of the butter and cheese. The walls of the dairy-house should be double, with an air-space, so as to preserve, as much as possible, the proper temperature, varying from 50° to 55° F. We would recommend hollow bricks for the walls of dairies. These are less liable to damp, from not being absorbent,—the air enclosed within them gives them this peculiarity,—and they retain a more equal temperature within the walls by impeding the transmission of heat.

In winter, it is equally important that the requisite temperature should be constantly maintained. If the building forms part of the house, it will generally be found sufficiently warm without the addition of artificial heat; but in very cold weather, and in detached dairies, unless they are constructed as already described, it will be difficult to preserve the proper temperature without the aid of a stove. In large dairies the expense would be of no consideration, when put in comparison with the advantage; but great attention is required in the control of temperature, for if too much warmth is generated, it will be as injurious as too little, and it will be altogether useless if neglected during the night, for if the dairy is once allowed to become too cold the injury done to the milk cannot be repaired by afterwards warming it. Probably the best way of warming a milk-room in winter is by a well arranged set of hot-water pipes, along the walls and near the floor.

As the greatest cleanliness is requisite in the various departments of the dairy, a well-arranged building should have separate divisions in order that its business may be properly performed. A *butter dairy* should comprise two distinct compartments, one for receiving the milk, another for performing the operation of churning, and, in addition thereto, a shed for washing utensils, and for the boiler. For a *cheese dairy*, three rooms will be requisite, viz., a milk-room, as before, for making the cheese, a second for salting and pressing it, and a third (which may be commodiously placed as a loft over the others) for storing and preserving the cheese until brought to market. An open shed formed by the projecting roof of the building will generally be found sufficient to scour and dry the vessels in. The dairy should be provided with a boiler, of dimensions suitable to the number of cows kept; and there should also be sufficient space for the convenient performance of all the operations of the dairy, whether it is devoted to the manufacture of butter or cheese.

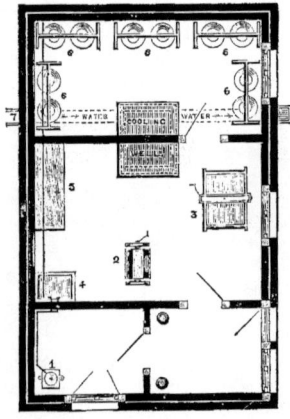
Fig. 53.—Butter Dairy—Ground Plan.

Fig. 54.—Model Dairy.

In fig. 53 we give the plan of a dairy, consisting of a milk-room, churning-room, and a third room divided off into boiler-room, and room for utensils. In the first are seen the milk-stands marked 6, and a fresh-air inlet at 7. In the middle room are respectively seen (2) the churn, (3) the butter-worker, (4) the washing-trough, and (5) the table. In the remaining room is indicated the position of (1) the boiler and hot water cistern. In fig. 54 we give an elevation of this dairy.

During recent years efforts have been made to produce an efficient machine for milking cows, but so far without the measure of success that would be widely welcomed. Cows are being milked by machinery, but not extensively. We may hope that success will yet reward the men who persevere.

The most important and wonderful appliance ever invented for use in a butter-dairy is the Centrifugal Cream Separator, which has been developed in an extraordinary manner since 1877, in which year we saw the germ or initial idea, in its embryo state, at the Hamburg International Dairy Show. The machine is pretty near perfection, and there are many sorts and sizes of it. It will get more cream out of the milk than can be got by any other means, and it is simple enough to be placed in the hands of anyone possessing the rudiments of mechanical knowledge.

In fig. 55 is seen the Dairy Supply Company's steam- or oil-power

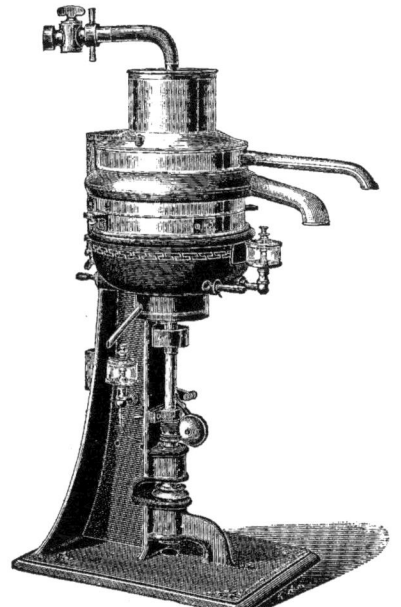

Fig. 55.—Steam- or Oil-power Laval Separator.

Fig. 55a.—Hand-power Laval Separator.

Laval Separator, the capacity of which is to separate the cream from 135 gallons of milk per hour, and fig. 55a shows one of their hand-power separators. The "Baby" is a smaller machine, and there are power machines varying in capacity up to 440 gallons per hour.

The well-known "Alexandra" separator, of which a sectional illustration was given in our fourteenth edition, has been displaced by a neat and compact machine named the "Lister," manufactured by Messrs. Lister & Co. of Dursley. Old types give place to new ones amongst separators, as amongst most other mundane productions, and this may be accounted as evidence of improvement—presumably so, at all events. It is true enough in respect to separators. Machines are now on the market which, by hand-power only, and that not too strenuous, effectively separate upwards of 100 gallons an hour, whilst power machines are efficient up to and above 400 gallons an hour. Fig. 56 shows a machine of the capacity of 50 to 100 gallons per hour.

The sterilization of milk has become quite a considerable branch of the dairy industry. Fig. 57 depicts the Simplex Sterilizer, introduced by the Dairy Supply Co. of London. This apparatus is made in various

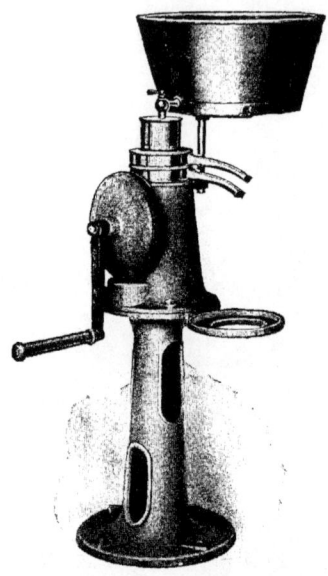

Fig. 56.—The Lister Separator.

sizes, from a capacity of sterilizing 90 to 880 gallons of milk per hour, while a form of different pattern deals with 50 to 1,200 gallons per hour. For the sale of milk in bottles, to be kept for an indefinite period, as on a voyage by sea for example, sterilization, of course, is essential; but a large quantity of milk is also sterilized or "pasteurized" for immediate consumption, as a safeguard against infection from germs of disease possibly contained in it.

Dairy UTENSILS comprise pails, sieves, coolers, churns, creaming-dishes, cheese-vats, ladders, presses, &c.; all of which are so familiar to every dairy-woman, that it would be only waste of time to describe them. To these should be added a Fahrenheit's thermometer, which should be suspended in a central part of the milk-house. Wood is the material usually employed as a frame in which the thermometer is set, but, even with the greatest care, wooden instruments are apt occasionally to be damp, and to acquire a faint musty smell; the closest attention in scouring and scalding every time they have been used is requisite, as the smallest drop of milk left in them, or the least taint of acidity or mustiness, may spoil the next milk. A metal thermometer-frame will be found more satisfactory, but many disadvantages are avoided by using simply a plain glass thermometer without frame or case.

In some dairies, wooden vessels lined with lead are used. Wherever

Fig. 57.—The Dairy Supply Co.'s Sterilizing Apparatus.

the size and shape of the utensil will admit, earthenware vessels properly glazed, or glass utensils, will be least troublesome, and glass, being so cheap, now places these latter within the means of most dairy farmers; but lead, copper, or brass utensils, as well as earthenware vessels glazed with lead, although found in many dairies, are to a certain extent objectionable, for the acid contained in milk that has been long exposed to the air forms an injurious compound with these metals, and this, although perhaps not deleterious to any serious degree, has occasionally been found to impart a disagreeable flavour to the milk. Cast-iron, although it does not form an absolutely poisonous compound with the milk, is by no means unexceptionable, because the result may, in a considerable degree, affect or change the taste and quality of dairy products. This, however, may be perfectly prevented by a due regard to cleanliness. The best milk-pans are of sheet-iron, enamelled inside, and seamless, or else of porcelain. Excellent dairy utensils may now be purchased almost anywhere.

A most convenient and useful milk-stand is seen in fig. 58; this stand may be unhesitatingly recommended as a room-economiser, and,

Fig. 58.—Revolving-disc Milk Stand.

as the discs on which the pans are placed revolve, skimming is greatly facilitated.

The late Dr. Voelcker, in speaking on the shape and size of milk-pans, said that " according to the experience of good dairymen, shallow vessels were the best. They threw up more cream, and preserved the milk better, which were very important considerations. Milk could not be kept together of any depth without its getting heated and spoilt. It was an erroneous view to take, to say that excess of air was injurious to milk. He would recommend that the air should be allowed to penetrate the milk and come in contact with it freely. If, too, they could maintain a current of air through the dairy, it would be all the better; but what would prove very injurious was to have the damp air resting upon the milk. Recently, a little work had been published in Sweden, which recommended that the milk should be exposed in shallow vessels of a peculiar shape, handy construction, and freely admitting the air. A part of the author's plan was to have a fire in the dairy whenever it was required; and he was informed that when a thunderstorm was seen approaching, instead of keeping the milk cool, a fire was at once lighted, and steam got up, to drive out the additional quantity of moisture. That might be a curious proceeding; but he could readily understand it. It was the damp, moist, heavy air that spoiled the milk. Remove that air by any means, and the milk would keep. It was of the utmost importance to have a dry air in the dairy;

Fig. 59.—"Charlemont Diaphragm" Churn.

and they could now understand why good dairymen always kept the floor as dry as possible. When a thunderstorm approached, the air generally became saturated with moisture, and that moisture had a great deal to do with spoiling the milk. If, however, they drove off the moisture, and with it the excess of water, the milk would keep; so that even in hot weather, when a thunderstorm occurred, an additional fire would preserve the milk good. The fact was a curious and instructive one."

Slate makes very good milk-coolers, and in some of the midland counties the common flag-slate is employed for this purpose. But, were it not for their fragility, glass and Wedgwood ware would be unrivalled.

Dairy utensils should always be first washed with cold water, then most carefully cleansed with hot water, and afterwards well rinsed with cold water, and kept in an airy place, in order that every possible trace of acidity may be removed.

The number and variety of churns made to-day is almost bewildering, and those by Bradford, Waide, Hathaway, Llewellyn, and other makers, are all excellent in their way, and as nearly as possible perfect. Bradfords' "Charlemont Diaphragm" churn, as seen in fig. 59,—at once simple in construction and easy to clean—is a churn which has, and probably will have, no superior. As will be seen in the illustration, the lid forms one end of the churn, and the large opening greatly

Fig. 60.—"Post Diaphragm" Churn.

facilitates the removal of the butter and the cleaning of the churn. The "diaphragm" beaters are easily removed, so that the churn has all the advantages of an "end-over-end" churn, and those of the time-honoured barrel churn as well.

The "Post Diaphragm" churn (fig. 60) is the latest development of this most effective principle of "diaphragm" beaters in churns, and it is believed that this particular form of churn will give the greatest effect. The churn—whose two sides are each composed of two inclines,—when rotating, carries the cream with accumulative force, in a wave of equal depth (in a round churn the volume of cream is greater in the centre than elsewhere) against the louvres of the diaphragm. And when it is considered that at the slow and easy speed of forty to forty-five revolutions per minute, the cream passes with this accumulative and increased concussive force eighty to ninety times per minute through the

Fig. 61.—" Cotswing " Churn.

louvres, its remarkable churning efficiency will be understood at once by those who have had experience in churning. The lid is seen on the bracing below. To this churn a drainer plug is fitted, and by means of it the butter-milk is drawn out, leaving all the butter in the churn; the plug is seen in the lower half of the churn.

The "Cotswing" churn (fig. 61), is one which simply oscillates,

Fig. 62.—The "Morning" Churn.

like a child's swing-cot, and the cream is churned by being thrown against each end of the churn alternately. It is, of course, extremely simple alike in construction and operation, and also effective; at the same time, save on the ground of fancy, it cannot be recommended in preference to the Diaphragm Churn. The "Morning" churn is a small churn, for two or three quarts of cream, adapted for persons who keep one or two cows only and want fresh butter frequently, or for those who wish to test the butter-yielding quality of different samples of cream. The illustration (fig. 62) shows a frame containing two separate vessels, and the churn is made with one, two, or three of them in a frame, the single one being also made with a dwarf frame, to stand on a table.

The Triangular Concussion Churns, free from beaters (fig. 63), manufactured by Messrs. W. Waide and Sons, Leeds, make butter on the usual principle of concussion, and, in working, must not be over-

Fig. 63.—Triangular Concussion Churn.

filled. They are made in various sizes to churn from two gallons to forty-five gallons.

At the Plymouth Meeting of the Royal Agricultural Society, in 1890, the Dairy Supply Company, Ld., exhibited the Instantaneous Butter Maker. The Laval Steam Turbine Separator is employed, and to this the new churn, invented by Dr. De Laval, of Sweden, is attached. It consists of a cylinder about 12 inches long and 4 inches in diameter, within which a dasher revolves at about 3,000 revolutions per minute, being driven by a rope belt, of the same kind as is used to drive a power separator, from the separator spindle. The cream, on leaving the separator in the usual way, passes over an ingeniously contrived refrigerator of new design, which reduces the temperature as low as possible with a very small consumption of cold water; it then enters at one end of the cylinder, in the course of its passage through which the cream is churned into butter, and emerges at the other end in a granular form. Dairymen who have had their butter-milk analysed from time to time, know that there is great loss in the present system of churning large quantities of cream, as it is impossible to ensure that every butter globule shall receive the same amount of concussion, and hence the butter-milk often contains a large percentage of butter; this is now avoided, as the cream must pass equally through the cylinder, at the same time receiving a regular and rapid concussion from the revolving dasher. The cylinder is enclosed in a water casing, so that the temperature is kept very low, and the butter is consequently firm, whilst there is no possibility of its being overchurned. It is very free from butter-milk, and therefore keeps well.

Fig. 64.—"Arch-Albany" Butter Worker, with Helical Roller.

It might be expected that this practice of churning fresh cream would entail a loss in the quantity of butter, but as a matter of fact it has been ascertained from experiments that this is compensated for by the perfect separation of the butter from the butter-milk. The churn is fixed to the separator frame, and can be attached to any of the Laval machines. As shown on the turbine, the whole process of separating the milk and churning the butter is performed by a jet of steam direct from the boiler, without the intervention of shafting, belting, or an engine of any kind. The churn has no complex arrangements about it, it can be taken to pieces and cleaned with the greatest ease, and it is an advantage that whilst the separating and churning can be done at one operation, yet they are independent of each other, so that the milk is separated at a warm temperature, ensuring the greatest yield of butter, and the cream is churned at a low temperature, ensuring the finest possible quality. The process is entirely automatic, requiring very little power and attention whilst in use. It is, therefore, a machine for saving labour in the dairy.

The end-over-end churns, free from beaters, and supplied by several

Fig. 65.—"Arch-Albany" Butter Worker, showing the Making-up Board in position, and the Roller when not in use.

firms, are favoured by a great many butter makers. The ancient "Dolly" churn is still in use in some places where butter is churned from milk.

There are various kinds of butter-workers, both for hand use and power, but the most efficient is the kind seen in figs. 64 and 65, this particular specimen being the latest development of the idea, and, as we can speak from experience, a most convenient and effective instrument. The roller, instead of being fluted longitudinally, is grooved helically, like the worm of a screw, the effect of which is that the butter-milk is expressed more certainly, and with less than one-half the rolling, and the granulation of the butter is preserved. The arched form of the table assists the moisture to flow away from the butter at both ends. The back action of the helical roller brings the butter back into a mass, ready for rolling out again. The "Arch-Albany" butter worker is made by Messrs. T. Bradford & Co., 140, High Holborn.

CHAPTER IV.

THE SECRETION OF MILK.

IN order adequately to understand the organic mechanism whereby milk is produced it is desirable first to inquire what becomes of the food which the cow eats—of the grass she grazes in the pasture, or of the cake, roots, or hay supplied to her in the stalls.

Cattle, sheep, and ruminants generally are popularly described as possessing four stomachs. It is, perhaps, more correct to say that these animals each have one stomach, comprising four compartments. The names of these are many ; in the order in which the food traverses them they are :—

1. The rumen or paunch ;
2. The reticulum or honeycomb ;
3. The omasum,[1] psalterium, liber, manyplies, manyplus, or many-leaves.
4. The abomasum, or reed, or rennet stomach.

The capacity of the stomach of the cow is enormous, amounting to from fifty to sixty gallons. It fills the greater part of the abdominal cavity, and the paunch alone occupies nine-tenths of the entire volume of the stomach, the remaining three divisions constituting a mere chain on the front left side of the paunch. In the sheep, though absolutely smaller, the paunch is relatively as large as in the ox. The fourth division, or abomasum, is the part of the cow's stomach the internal lining membrane of which secretes gastric juice. In other words, only the fourth compartment is capable of exercising the digestive function. It is called the rennet stomach, because it is the fourth compartment of the calf's stomach which is salted and preserved in the form of "vells" to furnish natural rennet for use in cheese-making. The secretion of the peptic glands, which line the abomasum, supplies the rennet.

Like all ruminants the cow can stow away in the rumen or paunch, as the first division of the ruminant stomach is called, an enormous

[1] Gr. *omos*, raw.

quantity of vegetable food. This, at a suitable time is regurgitated into the mouth, where it is mixed with abundant juice of the salivary glands and reduced to a fine condition between the teeth. Passing again down the gullet, the masticated food is this time directed into the fourth division of the stomach—the reed or rennet stomach, or abomasum. The glands lining this stomach pour out abundant gastric juice upon the food, which is at the same time kept in continual motion by the peristaltic contractions of the wall of the organ. Through a narrow aperture, the pylorus, the food, which is now called chyme,[1] passes next into the small intestine, a tube from half an inch to three quarters of an inch in diameter, and some fifty yards long. About two feet from its place of origin at the pylorus the small intestine is pierced by the bile-duct, a vessel which pours into the intestine the special secretion it derives from the liver. About fourteen or sixteen inches farther on another tube enters the small intestine, this is the pancreatic duct which conveys from the pancreas or sweetbread a juice which is likewise poured into the intestine. Thus the small intestine receives from outside itself two secreted fluids, the bile and the pancreatic juice, and furthermore the inner lining of the small intestine itself is beset with glands which pour out a juice called the succus entericus. Finally, the small intestine after coiling about in an indescribable manner opens abruptly into the side of the large intestine, a tube of varying diameter, and from thirty to forty feet long, its excretory orifice communicating with the exterior of the body.

Thus, the food taken in at the mouth is passed through the pharynx into the gullet, which leads into the stomach, where the food is reduced to a sort of pea-soup consistency, and is then called the chyme. This escapes through the pylorus into the small intestine, after traversing which the food material enters the large intestine terminating in the excretory orifice whereby the refuse of the food is ejected. During its passage along the alimentary canal the food is attacked by a number of digestive juices, comprising the saliva, the gastric juice, the bile, the pancreatic juice, the succus entericus, and the juice of the large intestine. Each of these juices has its own special and appropriate function, the gastric juice, for example, dissolving the nitrogenous constituents or proteids of the food, the bile assisting in emulsifying the fats, and the general result of their combined action being to separate from such apparently unpromising materials as grass or roots, hay or oil-cake, their nitrogenous or flesh-forming constituents, their carbohydrates or sugar-like ingredients, and their fats or oils. Hence, what is taken in at the mouth as hay or grass becomes, in the small intestine, a grumous mixture of soluble peptones, derived from the nitrogenous food constituents, and of soluble carbohydrates, emulsified fats, and indigestible fibre.

But the digestion of the proximate constituents of the food-stuffs in the intestine would be of little avail, did not the system provide some means whereby the contents of the intestine can be removed from that

[1] Gr. *chuma*, a thing poured.

tube, and transferred to any or every organ of the body where they may be required, no matter for what physiological purpose. Such means are afforded by the blood, and by the blood alone. It is an established fact that most of the absorption of nutrient substances contained in the food takes place from the chyme in the small intestine. To understand how this absorption is effected it is necessary to inquire into the structure of the small intestine and to see what facilities the arrangement of its tissues offers for the accomplishment of this purpose.

Of the several coats or layers which make up the small intestine the innermost one is of chief interest in this connection; it is called the mucous membrane of the intestine and it is beset with numerous small simple glands which secrete the succus entericus, but of which no further mention need be made. This internal lining membrane is also furnished with innumerable small outgrowths which impart to it a somewhat villous or velvety appearance, and each little process is appropriately called a villus. The villi are almost microscopic in size, and they are so abundant and close-set as to confer upon the free surface of the mucous membrane an appearance like that of the pile of velvet. The structure of a villus is somewhat complex; its outer part consists of a coat of delicate thin-walled cells forming what is called an epithelium. Along the middle of the inside space of the villus there extends a more or less branched thin-walled tube, the various parts of which originate blindly within the villus and coalesce into one main tube which passes out of the villus at its base and finds its way into the deeper walls of the intestinal canal. This narrow tube is similar in origin and structure to numerous other tubes which are to be found arising spontaneously in nearly all parts of the body, and are known as lymphatic capillaries, but, for a reason that will be presently mentioned, the lymphatics of the small intestine are distinguished under the name of lacteals. Between the lacteal and the epithelial wall of the villus there is a magnificent network of delicate, narrow, thin-walled blood-vessels (capillaries), and at the base of each villus a minute artery enters and breaks up into the capillaries which subsequently coalesce so that their contents are poured into one or two equally minute veins which leave the villus at its attached or basal end. It is not difficult then to imagine the structure of a villus : the central or axial part is occupied by the lacteal, this is quite surrounded by a network of blood capillaries, which, in its turn is completely enveloped by a sheath of epithelium, enclosing the inner structures like a thimble does the tip of the finger (fig. 66, page 274).

The importance of the villi will be appreciated when it is stated that it is through their agency that the nutrients of the food are abstracted from the chyme of the small intestine. Substances in solution, such as peptones, carbohydrates, and salts, pass readily through the epithelium of the villus and through the delicate walls of the blood capillaries into the blood itself, and so leave the villus by the little veins that pass away at the deeper end. Hence, the blood that leaves the walls of the intestine by the intestinal veins differs in composition from that

T

which is brought to them in the intestinal arteries; notably it has gained soluble nitrogenous compounds and soluble carbohydrates. All the blood collected from the intestinal walls is poured ultimately into a large vessel called the portal vein, which enters the liver and distributes its contents among the capillaries of that organ, so that the blood from the intestine is submitted to the action of the liver, though what that action is it is not necessary here to inquire. The blood of the liver is in the end

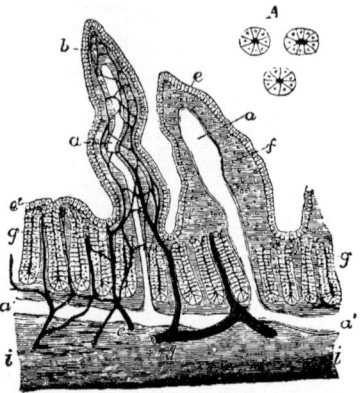

Fig. 66.—Vertical Section of the Mucous Membrane of the Small Intestine (Magnified 150 diameters).

Two villi are represented. In the one on the right hand the dilated lacteal alone is shown, in the other the blood capillaries and lacteal are both seen injected, the lacteal white, the blood-vessels dark; the section is carried through the tubular glands into the sub-mucous tissue; a, the lacteal vessels of the villi; a^1, the horizontal lacteal, which they join; b, capillary blood-vessels in one of the villi; c, small artery, conveying blood to the capillaries of the villus; d, vein, carrying blood away; e, the epithelium cells covering the villi; g, tubular glands (called crypts of Lieberkühn), which secrete the succus entericus, or intestinal juice; i, the sub-mucous layer. A, cross-section of three tubular glands more highly magnified.

collected by the hepatic veins, which pour their contents into a great vein called the posterior vena cava, and this passes forward and opens directly into the right side of the heart.

Let us return to the intestine and find out what becomes of the fats of the food. Very minute quantities of fatty matter may find their way into the blood capillaries of the villus, and so pass into the portal vein, but by far the greater quantity of the particles of emulsified fat pass into the villus and enter, not the blood capillaries, but the lacteal. It will presently be useful to inquire exactly how this is effected, but it is more to the point now to see what becomes of the fat after it enters the lacteal. The lymphatics of the small intestine resemble the blood capillaries in that they coalesce into larger tubes or vessels, and it is because, after a meal rich in fatty matters, the lymphatics of the small intestine are seen to contain a milky-looking fluid termed chyle,[1] that these tubes have received the name of lacteals. It may be well to mention here that this fluid is quite different and distinct from that which is obtained from the mammary gland. The lacteals which contain it find their way to the posterior end of a long irregular tube, which extends beneath the spine, and there discharge their contents. Because the greater part of the length of this tube extends along the dorsal side of the thorax, or chest, it is termed the thoracic duct, and the dilated posterior end which receives the emulsion of fat from the lacteals is appropriately named the receptacle of the chyle. The thoracic duct also receives the fluid collected by most of the lymphatics from all parts of the body, but this is a thin watery fluid consisting chiefly of the oozings through the walls of the blood capillaries. The thoracic duct must, of course, get rid of the materials it is continually receiving, and in the cow it pours its contents, about in the region of the first rib, into a large vein which joins almost immediately the anterior vena cava, another great vein opening into the right side of the heart. Thus it appears that the nutritious ingredients of the food, whether they leave the intestine by the blood capillaries of the villi, or by the lacteal roots of the same structures, find their way into the blood of the right side of the heart.

The tubes or vessels which convey blood to the heart are called veins, those which carry blood from the heart are arteries. The heart is a hollow muscle possessing two main cavities, one on each side, and so arranged that there is no lateral communication in the heart itself between these two chambers. The entire apparatus of the circulation is so constructed that the rhythmical contractions of the heart shall drive the blood in one direction, and in one direction only. When the heart contracts—and its contraction produces the "beat"—the blood is driven into certain arteries which break up into smaller arteries, and finally into exceedingly narrow tubes called capillaries, so that if a fleshy part of the body be cut without injuring any blood vessel discernible by the unaided eye, the blood nevertheless wells forth from the severed capillaries. The capillaries gradually coalesce into small veins, and these into large veins, while the largest veins of all pour their blood into the heart again. The blood on the right side of the heart differs from that on the left side; the former is dark, almost black, while the latter is bright scarlet. The reason for this may be discovered by examining somewhat more carefully the results of the heart's action.

Commencing with the blood in the right side of the heart, the effect of the systole or contraction of the organ is to drive the dark blood out through the pulmonary artery into the lungs, in the capillaries of which the hot dark fluid is exposed to the influence of atmospheric air. The blood receives from the air its oxygen gas, and gives in exchange carbonic acid gas and water vapour, which pass out in the expired air. It is this deprivation of carbonic acid gas and addition of oxygen gas which causes the dark blood to become scarlet. The blood in the capillaries

---

[1] Gr. *chulos*, juice.

of the lungs is collected by the pulmonary veins and poured by them into the left side of the heart, whence it is, in its oxygenised state, driven out through a strong artery, the aorta, which through its branches supplies the capillaries of all parts of the body except the lungs. One of its branches, for example, supplies the kidney, where the nitrogenous waste of the blood is separated in the urine, another supplies the mammary gland in the cells of which the milk is elaborated from the blood, various other ones break up into the innumerable capillaries beneath the skin which permit of their blood giving up a considerable quantity of moisture in the form of perspiration. But sooner or later all the capillaries of the body in general pour their blood into small veins, and these into larger veins, till finally all the blood so collected is poured into one or other of the great veins, the posterior vena cava and the anterior vena cava, which have already been spoken of, and from them the blood passes again into the right side of the heart.

The walls of the blood capillaries are so thin as to allow of their being permeated. Hence, it is in the vicinity of the capillaries that the nutritive work of the blood is performed, and new tissue is built up, and it is also by means of the capillaries that the waste materials that accumulate in the tissues of the body find a road into the blood and are carried away; were they allowed to accumulate in the tissues disease would speedily ensue. It is in these capillaries of the system in general that the blood, in consequence of the duties it discharges, becomes more laden with carbonic acid and changes colour from scarlet to black. Moreover, in the irrigation of the tissues, which is a necessary consequence of the oozing of the blood through the capillary walls for the performance of its nutritive work, some means are requisite to convey away the fluid which would otherwise accumulate, and these means are afforded by the minute tubes called the lymphatic capillaries which convey their watery contents, the lymph,[1] mostly into the thoracic duct, so that in the end the overflow from the blood capillaries finds its way again into the blood—into that, in fact, which enters the right side of the heart.

As the mammary gland can only elaborate milk out of the materials brought to it by the blood, it may be as well to mention the route which the blood takes in travelling from the heart to the udder. The arterial blood is pumped from the left side of the heart into the aorta, passing along which, the blood reaches the external iliac artery, and this is continued on into the femoral artery, extending more or less parallel to the femur, or thigh-bone. The femoral gives off a branch, the prepubic, which in turn gives off a branch, the external pudic, and this, after passing through the inguinal ring, divides into two branches, the anterior, or subcutaneous abdominal artery, and the posterior abdominal, or mammary artery, and it is from these that the blood supply of the capillaries of the mammary gland is immediately derived; of the two the mammary artery is the more voluminous.

The blood after passing through the capillaries of the mammary

Lat. *lympha*, water.

gland is collected into the abdominal subcutaneous vein, commonly known as the "milk vein." In cows this vessel is particularly large ; it extends along the under surface of the abdomen to near the end of the sternum, or breast-bone, where it turns inwards to join the internal thoracic, or internal mammary vein, the openings in the abdominal wall through which these vessels pass being known as the milk fountains or doors. The internal mammary conveys its blood to the vein of the arm, and this joins the anterior vena cava which empties into the right side of the heart. By this route, then, the blood which has been submitted to the action of the mammary gland is returned to the heart.

Although the nutrients of the food stuffs have been shown to enter the blood, it is not implied that they there preserve their individuality. Indeed, it is easy to show that the contrary is the case. Blood consists of a liquid plasma in which are suspended enormous numbers of microscopic solid bodies called corpuscles, and the red colour of the great majority of these imparts the characteristic tint to blood. Physically, milk resembles blood, in that it also consists of a watery fluid in which are suspended immense numbers of minute solid bodies, the fat globules, which, being white, make the whole milk appear to be this colour ; the different colour of skim-milk is partly due to the fact that most of the white fat globules have been removed. Blood is slightly heavier than milk, the specific gravity of the former being 1·055 and of the latter 1·03. Blood placed in contact with non-living matter speedily coagulates, milk in similar circumstances does not. The coagulation of the blood is due to the separation of a material called fibrin from the plasma, and the entanglement of the corpuscles in the meshes of the fibrin. Thus is formed the clot, and the clear pale liquid which remains after separation of the fibrin from the plasma is called the serum.[1] Hence the blood consists of serum, fibrin, and corpuscles, though the fibrin does not exist as such in the living blood. In round numbers, the percentage composition of the serum is, of water 90 ; of nitrogenous substances, 8 to 9 ; of fat, extractive, and saline matters, 2 to 1. Of the corpuscles there are two kinds, the red and the colourless, but the former are nearly a thousand times as numerous as the latter, and contain 56·5 per cent. of water, and 43·5 per cent. of solids, the latter being almost entirely nitrogenous organic matter. The fibrin which separates from the plasma is also made up of nitrogenous organic matter. When the corpuscles on the one hand and the serum on the other are dried and ignited, and their ashes analysed, the leading mineral constituents of the corpuscles are found to be the chloride and phosphate of potassium, and of the plasma soda and chloride of sodium. "The corpuscles differ chemically from the plasma, in containing a large proportion of the fats and phosphates, all the iron, and almost all the potash, of the blood ; while the plasma, on the other hand, contains by far the greater part of the chlorine and of the soda."[2] The extractives of the blood, though not abundant in quantity, are numerous and variable, the chief ones being urea, kreatin, sugar and lactic acid. These few details may serve to show what a very complex

[1] Lat. *serum*, the watery part ; in particular, the watery part of curdled milk, whey.
[2] Huxley, "Physiology," p. 72.

substance the blood is, and what is the nature of the materials from which the mammary gland has to elaborate the milk.

The average percentage composition of the whole milk of the cow is contrasted with that of skim-milk in the following table:—

	Whole Milk.	Skim-Milk.
Water	87·0	90·0
Albuminoids	4·0	3·7
Milk-Sugar	4·6	4·8
Fat	3·7	0·8
Ash	0·7	0.7
	100·0	100·0

The albuminoids, or nitrogenous compounds, are casein and albumin, the latter in ordinary cow's milk constituting not more than one-ninth of the total albuminoids; the ash consists of lime, potash, soda, magnesia, and iron, with phosphoric acid and chlorine. As may be inferred from a comparison of the foregoing tables, most of the fat is removed in the cream; theoretically all the fat should be removable in this way, and in the most efficient centrifugal separator the residue of fat which is left in the skim-milk is as little as 0·2 per cent. It is further evident that the liquid part of the milk, after separation of the fat globules, still retains all the milk-sugar and most of the albuminoids. It is worthy of note, too, that skim-milk contains the same percentage of water as the serum of blood.

The subjoined tables, quoted after Duclaux, throw additional light on the subject of the composition of cow's milk:—

COMPOSITION OF COW'S MILK AND OF SKIM-MILK.

	Whole Milk.	After Setting.		After Centrifugal Separator.	
		Skim-Milk.	Cream.	Skim-Milk.	Cream.
Water	87·25	89·70	58·63	90·73	29·54
Fat	3·50	0·77	35·00	0·46	66·67
Casein and albumin	3·90	4·02	2·75	3·31	1·22
Milk-sugar	4·60	4·74	3·12	4·73	2·17
Ash	0·75	0·77	0·50	0·77	0·40
	100·00	100·00	100·00	100·00	100·00

PERCENTAGES OF DRY MATTER IN THE MILK OF THE SAME COW AT DIFFERENT DATES.

	August 11.		August 24.		September 28.	
	In Suspension.	In Solution.	In Suspension.	In Solution.	In Suspension.	In Solution.
Fat	3·22	...	2·75	...	2·34	...
Milk-sugar	...	4·98	...	5·38	...	5·07
Casein[1]	3·31	0·84	2·72	0·55	3·22	0·68
Phosphate of lime	0·22	0·14	0·21	0·14	0·18	0·22
Soluble salts		0·39	...	0·35	...	0·38
	6·75	6·35	5·68	6·42	5·74	6·35
Total dry matter	13·10		12·10		12·09	

[1] The casein "in solution" represents albumin.

COMPOSITION OF THE ASH OF COW'S MILK.

Chloride of sodium	16·23
Chloride of potassium	9·49
Potash	23·77
Lime	17·31
Magnesia	1·90
Oxide of iron	0·33
Phosphoric acid	29·13
Sulphuric acid	1·15
Silica	0·09
	99·40

In what way do the tissues of the mammary gland prepare milk from the blood which comes from the heart? Milk, like bile, gastric juice, pancreatic juice, saliva, and urine, is a secretion, and it is formed through the activity of certain living cells which constitute the internal

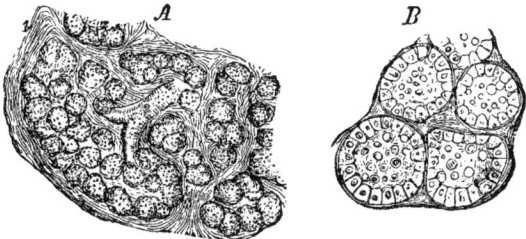

Fig. 67.—Magnified Sectional Views of the Mammary Gland during the Period of Lactation.

A, Section of a small lobule of the gland, magnified 60 diameters; 1, ground-work of connective tissue supporting the glandular tissue; 2, terminal branchlet or one of the excretory tubes; 3, alveoli.

B, Four alveoli, magnified 200 diameters, showing the lining epithelial cells and some milk globules.

lining of the ultimate branches or sacs of the mammary gland. Each terminal branch, in fact, is formed by the confluence of several blind, saccular or flask-shaped, wavy tubes, called alveoli[1] (fig. 67).

The cow's udder, or milk-bag, is provided with four delivery tubes or teats, each of which (fig. 68), with its gland, is termed a "quarter." When a cow is said to have "lost a quarter," it means that one of the teats has ceased to yield milk. Besides the external covering which binds together the whole of the udder, each gland has its own special fibrous envelope, and is distinct from, and independent of, the other glands; hence, though the function of one gland, or "quarter" may be impaired, the others may continue to act in the usual way. The orifice at the free end of the teat is a narrow tube, which is ordinarily closed. In the body of the teat this tube is much wider, but becomes constricted again at the region where the teat merges into the udder. Above the constriction is a large space, "the milk cistern," or

[1] Lat. *alveolus*, a little hollow.

reservoir, which becomes distended with milk as the secretion accumulates. Into each of the four milk cisterns innumerable tubes open. Any one of these may be traced back into minute tubes or ducts, which end blindly in several small sacs or bags, and it is these latter

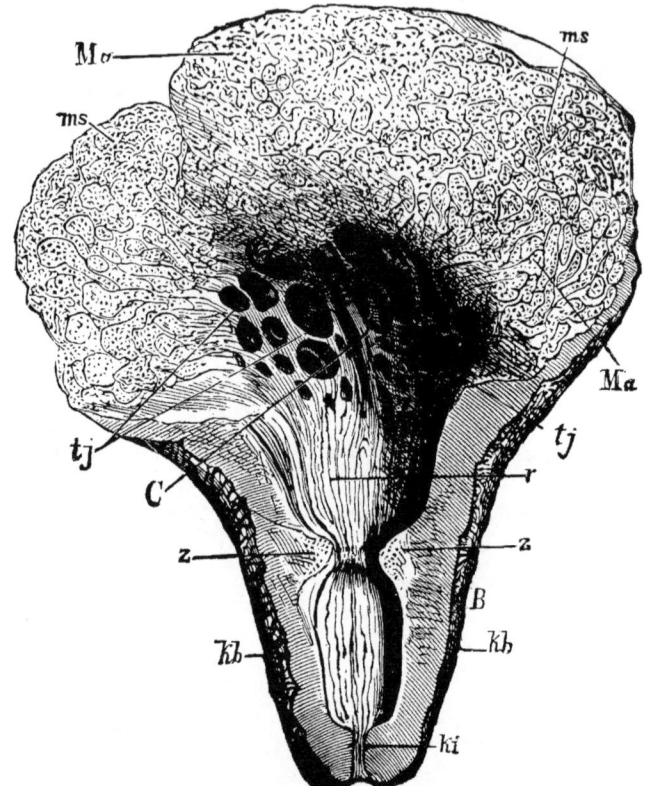

Fig. 68.[1]—Section of Udder and Teat of Cow (*Thanhoffer*).

Ma, gland substance; B, nipple or teat; *ms*, acini of gland; *tj*, milk ducts; C, milk-cistern; *r*, folds in wide milk-ducts; *z*, section of sphincter muscle; *kb*, external skin; *ki*, narrow milk-duct in the nipple.

which are called alveoli. The delicate walls of the ducts and the alveoli are lined by a single layer of minute living cells, which are the secreting cells of the mammary gland. The whole gland is richly supplied with blood by means of thin-walled blood capillaries, a dense

[1] Figs. 68, 69, and 70 are reproduced by permission from Dr. Meade Smith's "Physiology of the Domestic Animals." London and Philadelphia: F. A. Davis.

network of which surrounds every alveolus. Out of the blood thus placed at their disposal the secreting cells manufacture milk, which flows along the ducts, and accumulates in the milk cistern at the top of each teat. The general plan, here described, upon which the mammary glands are constructed, is similar to that of the salivary glands of the mouth.

In the active gland each alveolus encloses a relatively large cavity, varying in size in different alveoli; it is lined internally by a single layer of columnar epithelium cells, each containing protoplasm and a nucleus. The transparent homogeneous granular-looking substance called *protoplasm* is living matter, by and from which all the tissues of animals, and of plants, are built up; its composition is ever changing, it is throughout active life continually receiving new matter, and is constantly parting with material which has been elaborated within itself, hence it is impossible to assign to it any definite chemical constitution. "For the living and life-giving protoplasm is endowed with internal forces, and as the result of this, with an internal and external variability which is wanting in every other known structure; its active molecular forces cannot, in short, be compared with those of any other substance. The capacity which protoplasm has, in consequence of the forces which become manifested in it, of assuming definite external forms, and of varying these, as well as its capacity of secreting substances of different chemical and physical properties according to definite laws, is the immediate cause of cell-formation and of every process of organic life." [1]

Each epithelial cell of the alveoli of the mammary gland is capable of forming in its interior one or more oil globules of varying size. These may and generally do run together, and, pressing the nucleus—which is a somewhat denser aggregation of the internal protoplasm—on one side of the cell, give to the latter the appearance of a fat cell.[2] And a fat cell it would probably become, and the whole mammary gland would in all likelihood undergo a conversion into a mass of adipose tissue, if the fat were not ejected from the epithelial cells, and particularly if, in these circumstances, the production of fat were exalted at the expense of the production of casein or of milk-sugar.[3] The fat can be seen to be gathered in the epithelium cell, and to be ejected by the cell protoplasm through the wall of the cell into the cavity of the alveolus. The cell resumes its former solid character, and begins again to form oil globules in its protoplasm, and the epithelial cells, so long as the secretion of milk is continuous, go on again and again forming globules of fat without being themselves destroyed.[4] Thus, the fat of milk, whose myriad globules have become confluent in every pat of butter, is formed in the microscopic epithelial cell through the metabolism [5] of its protoplasm. The fat-globules as brought into

[1] Sachs, "Text-book of Botany, Morphological and Physiological."
[2] Klein, "Elements of Histology."
[3] Foster, "Physiology."
[4] Langer, quoted by Klein, "Histology."
[5] Gr. *metabole*, change.

view in the microscopical examination of normal milk are seen to vary considerably in size, the largest ones being several times as big as an individual secreting cell from the epithelial lining of the alveolus. These large globules are produced by the fusion of the smaller ones after their expulsion from the alveoli, and during their passage along the lactiferous ducts.

The milk-sugar, or lactose, of the milk is, like the fat, also a product of the metabolic activity of the protoplasm of the secreting cells of the mammary gland. Of the exact mode whereby the protoplasm of the epithelium cell elaborates milk-sugar from the constituents of the blood, little or nothing is known, but that the formation of milk-sugar is effected by these cells is proved by the fact that this particular form of sugar occurs nowhere else in the animal body, though grape-sugar, or dextrose, is a normal constituent of blood, chyle, and lymph. Another proof is afforded by the circumstance, that although milk-sugar is a typical carbohydrate, its occurrence in the milk is not dependent on the presence of carbohydrates in the food, for it is maintained in abundance in the milk of carnivorous animals when these are fed exclusively on meat, a nitrogenous food as free as possible from any kind of sugar or other carbohydrate. Of all the constituents of milk, the milk-sugar is least influenced by external conditions.

With regard to the casein of the milk, is this also a product of the metabolic activity of the protoplasm of the secreting cell, that is, is it manufactured in and by this protoplasm, or is it simply separated from the blood? Here, again, evidence points to the former as the correct interpretation of the origin of the casein, for when the action of the secreting cells is imperfect, as at the beginning and at the end of lactation, the albumin which normally is less than one-seventh of the casein is actually in excess of it, and albumin—that particular variety known as serum-albumin—is a normal constituent both of blood and of milk. But when the secreting cells are in full activity, the casein comes prominently forward as the leading nitrogenous constituent of milk.

Certain physical peculiarities are associated with the presence of casein in milk. As has been demonstrated by Duclaux, casein is found in milk in two forms. It is partly in suspension, and on account of its different specific gravity this sinks to the bottom of a vessel of fresh milk left at rest. It is also, on the other hand, partly in a kind of gelatinised condition, in which state it remains diffused through the milk. These two forms of casein pass insensibly the one into the other, and although it is possible by rigorous methods to distinguish between them, yet there is no fundamental difference. That which is described as being in a gelatinous semi-liquid condition is, indeed, as much in suspension as is the other form, the proof of which is that if milk is filtered, not through a paper filter the pores of which are large enough to permit the passage even of granules of butter-fat, but across unglazed porcelain, the two forms of casein are both arrested, and appear as a more or less coherent mass of gelatinous matter.

If to the liquid which passes through the porcelain filter a little acid be added, a white granular substance is separated which is still casein. Filtered through filter-paper and the clear liquid heated, another precipitate (albumin) makes its appearance. This albumin is regarded, however, by Duclaux, as simply a physical modification of casein. Milk contains, in fact, of albuminoid matter properly so called, only casein. But this casein exists really in three states,—in a state of perfect solution, capable of passing through a porcelain filter; in a state of mucous coagulation, uniformly diffused throughout the liquid; in a state of suspension from which when milk is left at rest it falls to the bottom of the liquid. For the sake of simplicity, and also because it is difficult to separately determine the quantities of the two last-named forms, Duclaux gives to these two collectively the name of solid casein, and he calls that which passes through the porcelain filter dissolved casein.

In normal samples of milk the proportion of solid casein is liable to variation, but the percentage of dissolved casein is always approximately the same. The weather, the temperature, the addition of water, the action of acids or alkalies employed in feeble proportions, of salts, &c., affect the latter little or not at all. The same is true of the action of rennet. It evades consequently all the operations to which milk is submitted in order to obtain therefrom its nutrient ingredients. In the manufacture of cheese, only the gelatinous casein and the casein in suspension are utilised. The action of rennet is to cause gelatinous casein to pass into the state of suspended casein,— that is, of curd easily separable from the serum or whey.

Experiments made by Thierfelder [1] lead him to believe that milk-sugar is produced from blood serum by the agency of a ferment which he calls saccharogen, though he has not succeeded in isolating this substance. Casein he regards as most probably formed from serum-albumin by an analogous ferment likewise present in the mammary gland.

Hence, we learn that the mammary gland, by the direct metabolic activity of its secreting cells, appears capable of forming, out of its protoplasm, typical representatives of the three great classes of food-stuffs, (1) proteids, albuminoids, or nitrogenous organic compounds represented by the casein and albumin, (2) carbohydrates represented by the milk-sugar, (3) fats represented by the oil globules. In other words, the secretion of milk may be regarded as "a process of moulting of the epithelial cells, which undergo decomposition, and discharge the resulting products into the excretory ducts." But, in order to discharge this complex function, the protoplasm must be nourished; wonderfully capable as it is, it yet would be powerless to do the work which living matter alone can do were it not furnished with the material with which to operate. And this material is abundantly, almost lavishly, supplied to

[1] "Biedermann's Centralblatt für Agricultur-Chemie," 1884.

it by the blood, for every alveolus in the mammary gland is surrounded by a dense network of blood capillaries, and the exceedingly thin walls of these as well as of the secreting cells are easily permeable by fluid. As has already been indicated, the blood is dependent for its nourishment upon the food.

Reverting now to the formation of fat by the secreting cells of the mammary gland, the mechanism of this process may perhaps be better understood by studying analogous, though not strictly similar, processes which are observed to occur, firstly, in a certain low form of animal life, named Amœba, and, secondly, in the epithelial cells of the intestinal villi during the absorption or appropriation of fat. Amœba is a minute microscopic organism which may be found in stagnant water, or in aqueous infusions of animal matter. It has very much the appearance of a particle of jelly, it consists almost entirely of protoplasm, and is continually changing its form. A nucleus is often present. The gelatinous body of the amœba can hardly be regarded as possessing a distinct external covering, though the circumferential part does differ in some minor details from the deeper-lying portions. This limitary layer has been aptly compared to the wall of a soap-bubble, which, though fluid, has a certain cohesion which not only enables its particles to hold together and form a continuous sheet, but permits a rod to be passed into or through the bubble without bursting it; the walls closing together, and recovering their continuity as soon as the rod is drawn away. In a similar way the amœba feeds, taking in and passing out solid matter, though the animal possesses no aperture ; the solid body passes through the outer wall, which immediately closes up and repairs the rent. Thus does the amœba take in the small, usually vegetable, organisms, which serve it as food, and subsequently get rid of the undigested solid parts.[1] From the food thus obtained the organism can make new protoplasm, and produce other organisms like itself. It lives, moves, eats, grows, and after a time dies, having been during its whole life hardly anything more than a minute lump of protoplasm. Certain substances serving as food are received into the body of the amœba, and there in large measure dissolved. The dissolved portions are subsequently converted from dead food into new living protoplasm, and become part and parcel of the substance of the organism. Simultaneously there is going on an ejection of old material, for the protoplasm is incessantly undergoing chemical change (metabolism), room being made for the new protoplasm by the breaking up of the old protoplasm into products which are cast out of the body and got rid of. These products of metabolic action have, in many cases at all events, subsidiary uses. Some probably serve to dissolve the raw food introduced into the amœba, and remain in its body for some little time for this purpose. Such products are generally called secretions ; others which pass more rapidly away are spoken of as excretions. The distinction between the two is unimportant and frequently accidental.[2]

In the case of the intestinal villi, it is absolutely certain that the

---

[1] Huxley, "Practical Biology."
[2] Foster, "Physiology."

finely divided fat does pass from the intestine, through the epithelial cells which envelope the villus, and so into the channel of the lacteal. Most observers agree that after a meal the epithelium cells of the villus are gorged with fat, the particles of which must have entered the cells very much as foreign particles enter the body of an amœba. The cells may thus be said to eat the fat, and subsequently to pass it on in the direction of the channel of the lacteal. There would thus be a stream of fatty particles through the cell, a stream in the causation of which the cell took an active part. In fact, under this view, absorption by the cell might be regarded as a sort of inverted secretion, the cell taking much material from the chyme and secreting it with little or no change into the villus.[1]

Professor E. A. Schäfer has actually demonstrated in his paper, "On the origin of the proteids of the chyle and the transference of food materials from the intestine into the lacteals,"[2] that lymph corpuscles play an important part as carriers of fat into the lacteals of the villi. They take up, amœba-fashion, the fatty particles from the epithelium cells of the villi, and thus fat-laden they wander towards the centre of the villus and enter the lacteal, inside which the coat of the migratory lymph cell is dissolved and its contained fat set free. It is particularly worthy of note that migratory or lymph corpuscles occur in the tissue between the alveoli of the mammary gland.

Considerations such as the foregoing lead to the splendid generalisation that the epithelial cells of the villi, the secreting cells of the mammary gland, and the colourless corpuscles of the blood, may be regarded as amœbæ,—that, in fact, the whole animal body may be viewed as groups of amœbæ, associated together for many and varied objects, the different groups exhibiting specialisation of structure in accordance with the nature of the work they are respectively called upon to perform, the functions of one group differing from those of another in conformity with the strict physiological division of labour which is an inevitable condition of existence, if the aggregation of cells is to exhibit any characters of a higher order than those which belong to the cell as an individual.

A comparison of the epithelial cells of the villus with those of the alveoli of the mammary gland seems to bring under notice a similarity of behaviour with respect to the disposal of the particles of fat with which they, in each case, become laden. In the former case the globules are ejected into the interior of the villus; in the latter case, they are ejected into the cavity (or lumen, as it is called) of the alveolus, and in both cases the mechanical action is comparable with that whereby an amœba ejects from its body such matters as it has no further use for. In all other respects, however, the epithelial cells of the mammary gland possess functions of a far more exalted order than those of the epithelial cells of the villi, for the former actually manufacture the fat they contain out of the blood they

---

[1] Foster, "Physiology."
[2] "Proc. Roy. Soc.," No. 235, 1885.

receive, while the latter apparently take up ready-made fat and pass it on. Moreover, the mammary secreting cells are further capable of elaborating milk-sugar and casein out of the constituents of the blood.

At the beginning and at the end of lactation, and equally during the suspension of that function, the mammary gland presents features which well deserve study. The resting-gland, that is, the gland of a non-pregnant or non-suckling animal, contains fewer alveoli than the active gland, but a great deal of fibrous connective tissue. The alveoli, too, are at this period solid cylinders with no internal lumen, but during pregnancy these solid alveoli rapidly multiply, lengthen, and thicken, owing to the division of the epithelial cells. When milk secretion commences, the cells occupying the central part of the alveolus undergo the fatty degeneration, and are at once excreted. It is the presence of these cells which imparts to the milk, for several days after calving, the peculiar properties in virtue of which it is called colostrum, or colostrum (figs. 69 and 70). The central cells of the alveoli are appropriately termed colostrum corpuscles, and their elimination provides the cavity, or lumen, inside each alveolus, into which the fat globules formed in the peripheral epithelial cells are ejected. The peripheral cells, it is to be noticed, do not sacrifice their position like the central ones, or there would be none left for the work of secretion; but Schmid asserts that even these finally break up, one by one, and are replaced by new epithelial cells derived from the multiplication of the other still active ones. The small bits of granular substance met with here and there in milk are the remains of the worn out and broken down protoplasm of such epithelial cells.[1]

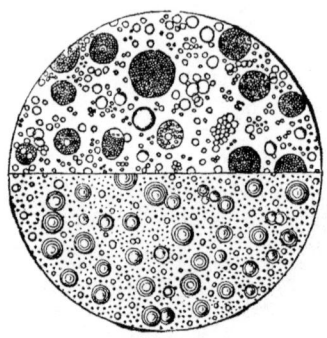

Fig. 69. — Microscopic appearance of Milk and Colostrum (*Landois*).

The lower half of the figure represents milk; the upper half colostrum.

The deep yellow colour, unctuous character, and higher specific gravity of colostrum are thus readily explicable. Fürstenberg thought that the pieces of membrane and clusters of cells which appear in this early mammary fluid were milk globules in a state of transition, that is, not yet perfectly formed.[2] But now they are known to be the degraded fat-laden cells of the axial region of the alveolus, which must, as it were, be cleared for action before the mammary gland can attain its full secretory power. The reason, too, that milk drawn within the first week after calving should never be used for making cheese is that such milk contains in relative abundance actual animal matter derived

[1] Klein, " Elements of Histology."
[2] Sheldon's " Dairy Farming."

from the débris of the central cells of the alveoli, and the presence of this animal matter must seriously interfere with the ripening of the cheese. The percentage of this material is, of course, largest at first, and steadily diminishes.

Colostrum is defined by Dr. Meade Smith, in his "Physiology of the Domestic Animals," as an opaque, yellowish fluid, containing a large amount of the so-called colostrum cells (true glandular cells in different stages of fatty degeneration), few milk globules, a large amount of albumin, little or no casein, and but little fat, milk-sugar, and salts. On account of its large percentage of albumin it coagulates when

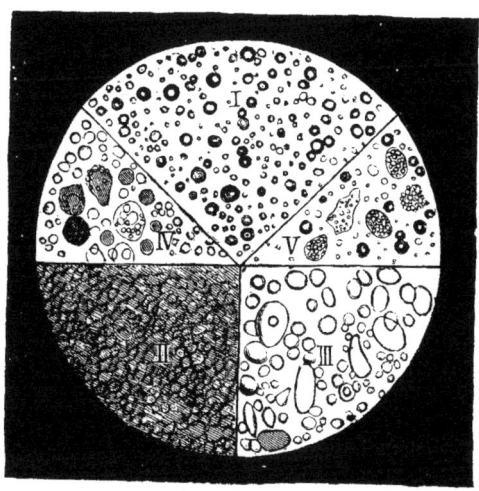

Fig. 70.—Microscopic appearance of (i) Milk, (ii) Cream, (iii) Butter, (iv) Colostrum of Mare, (v) Colostrum of Cow (*Thanhoffer*).

heated, differing in this respect from milk; in fact, colostrum when first secreted closely resembles blood-serum, with the addition of colostrum corpuscles. Gradually, however, the colostrum secretion passes into a milk secretion; albumin and colostrum corpuscles become reduced in quantity; fat, casein, sugar, and milk-globules increase. The specific gravity of colostrum varies from 1040 to 1060, being higher immediately after delivery, and falling as it gives place to a true milk secretion. The reaction of colostrum is ordinarily alkaline, and becomes acid on standing. Immediately after calving, the colostrum of the cow contains 8·5 per cent. of albumin, after one day 6·4 per cent., after three days 3·4 per cent., after seven days 1·9 per cent., and after twenty-one days 0·6 per cent. On an average colostrum may be said to have the following composition:—

Water	78·7
Casein	7·3
Albumin	7·5
Fats	4·0
Sugar	1·5
Salts	1·0
	100·0

The source of the water of milk must undoubtedly be sought in the water of the blood. As has previously been stated, the sanguineous fluid oozes through the finest blood capillaries in all parts of the body, and in most parts of the system this overflow is kept under control by means of the lymphatic vessels which collect the thin fluid, now known as lymph, and return it to the heart, though in most cases the lymph passes through the thoracic duct on its way to the central organ of circulation. The thoracic duct, it will be remembered, also receives the chyle collected by the lacteals, or lymphatics of the intestine, and it is estimated that in the course of twenty-four hours a quantity of fluid equal to that of the blood is thus poured into the duct on its way to the heart, about one-half of this fluid being lymph and the remainder chyle. Lymph is a pale yellowish fluid, containing from 94 to 95 per cent. of water, and may be regarded as blood deprived of its red corpuscles and diluted with water, while chyle may be described as lymph containing much fatty matter. The great distension which the mammary gland undergoes, as it becomes stocked with milk, is chiefly due to the accumulation of water, the presence of which is necessitated as a vehicle whereby the other constituents of the milk may be conveniently taken up and carried away. The copious supply of blood, in the capillaries which surround each alveolus of the mammary gland, is most favourably circumstanced for giving up its aqueous material through the moist delicate membranes by which it is enclosed; any watery fluid which did not go to swell the excretion of milk would find its way amongst the tissues constituting the framework of the gland into the lymphatics, and so back by way of the thoracic duct into the blood. The kidney affords another example of an organ whose secretion, mainly nitrogenous in this case, is mingled with much water and converted into an excretion.

It is important to remember that the mammary gland of the cow, particularly in such favourite milking breeds as the Jerseys, the Ayrshires, and the Dutch, has been brought into a condition of abnormal activity differing as much, perhaps, from the gland in its unimproved condition, as does the large shapely root of a cultivated turnip from that of its wild progenitor. The duration of lactation has been extended, the quantity of the lacteal fluid has increased. Moreover, the quality of the milk is found, within certain limits, to vary under different conditions, the percentage of water sinking at times to as low as 84, and rising to as much as 90. It is a legitimate and useful object of inquiry as to how far this kind of variation can be controlled, and particularly as to what extent it can be modified with economical advantage.

The morning milk usually contains from ½ to 1 per cent. more water than the evening milk, but on the other hand there is generally more of it. This is attributed to the fall of temperature during the night, which would necessitate the oxidation of more carbonaceous material to keep up the body warmth, and to the lessened activity of the animal. But Fleischmann and Vieth[1] experimenting on Count von Schieffen's Mecklenburg herd of 119 cows, found not only that the proportion of fat in the evening oscillated within wider limits than in the morning, but they also observed that from March to July, the period of greatest activity of the mammary gland, the morning milk was richer in fats than that of the evening, the cows being pastured on common land at the time. It is interesting to note, too, that the annual yield of milk of this herd was, for each cow, 5·69 times its own live weight. That the fat, though represented by a smaller percentage in the morning milk, may yet be present in greater absolute quantity is shown by the following average results obtained by the same experimenters[2] after a year's examination of the milk of several cows:—

	Morning milk.	Evening milk.
Specific gravity	1·0316	1·0318
Percentage of fat	3·374	3·420
Yield per cow in pounds	7·814	7·566
Yield of fat in ounces	4·224	4·083

The differences in the milk drawn at the beginning and at the end of milking have been studied by Schmidt,[3] who employed a middle-aged Dutch cow giving twenty-one imperial pints daily. A measured quantity (seven-eighths of a pint) was taken from the two hinder teats at the commencement of the morning milking, and an equal quantity at the end of the same milking. One hundred parts by weight of the milk contained:—

		First milk.	Last milk.
Total solids		9·20	13·64
Albuminoids	Casein	2·24	2·11
	Albumin	0·31	0·29
	Peptone	0·10	0·12
Fat		0·76	5·60
Sugar		5·08	4·92
Ash		0·69	0·66

Thus, the difference in the quantity of the total solids is almost entirely in the fat, which is between seven and eight times more abundant in the last than in the first milk. The explanation offered is, that the fat at first lodges in or adheres to the lactiferous ducts, and that a separation of cream begins in the udder, and this would, as far as circumstances permitted, seek to float on the denser fluid aggregated at the base of the teats. The udder of a cow killed immediately after milking showed, on examination, that the ducts contained a residue of rich milk. Excepting the fat, the great bulk of the milk secreted by a

---

[1] "Landwirthschaftliche Versuchs-Stationen," vol. xxiv.
[2] "Bieder. Centr.," 1880.
[3] *Ibid.*, 1883.

cow appears to possess a tolerably uniform composition, but the whole of the fat is rarely obtained in the milking.

Though much locomotion is detrimental to the yield of milk, it is a mistake to suppose that uninterrupted confinement in the stall is the most economical treatment for a milch cow. With moderate locomotory exercise the slight reduction in quantity of milk appears to be fully compensated for by the increased yields of solids. Munk,[1] to settle this point, experimented with thirty cows, and found that when they were allowed half an hour's daily exercise, the total quantity of the milk as well as the fat and casein increased, though much exercise exerted an adverse influence on the yield. When cows are on grass, their increased appetite in the presence of abundance of food quite makes up for any loss incurred in the movements necessary to obtain that food. Hence it is desirable that stall-fed milch cows should have daily exercise.

It is, in a sense, due to a mere physiological accident that the fat elaborated in the organs of a milch cow is excreted rather than stored up in the system. And the metabolic activity of the secreting cells of the mammary gland is closely connected with the exercise of the maternal functions. We are utterly in the dark, says Dr. M. Foster, as to why the uterus, after remaining apparently perfectly quiescent for months, is suddenly thrown into action, and within, it may be a few hours, gets rid of the burden it has borne with such tolerance for so long a time. After birth, the maternal energy, previously occupied in nourishing the fœtus, is now directed to the secretion of milk, whereby the nourishment of the offspring may still be maintained.

Though it may be physiologically possible for a cow to give a copious flow of milk, and simultaneously to undergo a marked increase in weight, nevertheless, these two effects are generally in an inverse ratio to each other. Thus, it is a characteristic of individual cows, and in some cases of entire breeds, as the Herefords, the Galloways, and the Aberdeen Polls, to lay on flesh rather than to make milk. Of the causes which determine such idiosyncrasies little or nothing is known, just as little or nothing is known as to why the same food, given freely to a wether and to a steer, should be converted into mutton in the one case, and into beef in the other. It is the duty of the breeder to discover these and other idiosyncrasies, and to endeavour to perpetuate them, if they appear to be desirable. In an insufficiently fed or starving animal, the first demand the system makes is on the fat stored up within it, so that a starving ox feeds as much on animal matter as does a lion. Similarly, in a milch cow, the first effect of insufficiency of food is the falling off in the amount of fat secreted by the mammary gland. Nor can this secretion become at all marked in quantity till all the other physiological requirements of the body have been first attended to. This was very well exemplified in the case of a two-and-a-half year old Southdown ewe experimented upon by Weiske.[2] The ewe lambed on April 22, and was regularly milked three times daily, receiving at the

[1] "Bieder. Centr.," 1884.
[2] "Bieder. Centr.," 1880.

time 1 lb. hay, and 1 lb. oats per day together with turnips. The weight of milk she gave each day from the beginning of May was:—

Date	1st	2nd	3rd	4th	5th	6th	7th	8th	9th
Grams	523	620	736	768	840	910	924	992	987

From May 10 to May 20 there was a very regular daily yield of 1,000 grams. On May 21 the ewe was shorn, and, notwithstanding the continuance of the same food and treatment, the following falling off in yield was recorded:—

Date	20th	21st	22nd	23rd	24th	25th
Grams	1006	913	854	781	750	712

The meaning of this is obvious. The removal of the wool necessitated the immediate consumption of more food to maintain the body warmth; the rations served out to the animal remained the same, and consequently the production of fat had to suffer, and this loss fell on the secretion of the mammary gland. On May 26, however, $\frac{1}{2}$ lb. of linseed cake was added to the other food, and the following yields of milk were measured:—

Date	26th	27th	28th	29th	30th	31st
Grams	687	760	889	950	910	961

Thus the decrease in the yield which resulted from the shearing, was made good by adding an appropriate quantity of linseed cake to the daily food of the ewe.

To some extent in this country, but more especially in countries where, as in Canada, the contrast between summer and winter is more marked, an error is made in the underfeeding of cows that are not in milk. Exposure to the winter's cold causes a larger demand to be made upon the food for its heat-giving properties, so that with the same amount and quality of food, less milk can be expected in winter than in summer; this necessity for increased food in colder weather may, however, be largely met by warm and comfortable housing. It is a blunder, both economically and physiologically, to imagine that if cows are just kept alive through the winter, all that is necessary has been done. On the other hand they need to acquire in winter a reserve of health and of flesh, which they can fall back upon in summer, when the flow of milk should be most copious. The frequent drying up of pastures towards the end of summer must be met by an extra supply of food, otherwise the milk-secreting function will be much impaired, for, as has been shown, it is quite subservient to other demands of the body.

Related in some degree to the change from winter to summer feeding is the not yet thoroughly understood occurrence of "lazy" or "heavy" milk, so called because of the apparent reluctance with which the cream separates. As complaints of lazy milk are generally most numerous at about the time when the cows begin grazing, Schrodt and Du Roi,[1] imagining that the sudden change from the dry hard feeding of the

[1] "Bieder. Centr.," 1880.

stalls to the tender soft herbage had something to do with the complaint, caused a herd of milch cows to be fed with green food previous to their going out on grass, commencing with small quantities of green fodder, and gradually increasing this until it had wholly replaced the other. The experiment extended over a fortnight, and at its close the animals were put out to grass, but there were no complaints of lazy milk. Hence, the abrupt change from dry to green fodder appears sometimes to result in lazy milk, but why this should be so is not known. Generally speaking, sudden and abrupt changes from one kind of feeding to another should always be avoided, not on account of the milk only, but for general considerations relating to health.

It might be supposed that the more fat a food was found, on analysis to contain, the richer would be the milk resulting from the consumption of that food. But this is a generalisation not altogether warranted by facts. Foster[1] significantly observes that the quantity of fat present in milk is largely and directly increased by proteid (*i.e.*, nitrogenous) food; but not increased, on the contrary, diminished, by fatty food. The explanation of this is, that proteid food increases, and fatty food diminishes, the metabolism of the body; in other words, that it is the nitrogenous constituents of the food-stuffs which excite the activity of the living cells on whose protoplasm the maintenance of the functions of the body is dependent. A bitch fed on meat for a given period gave off more fat in her milk than she could possibly have taken in her food, and that too while she was gaining in weight, so that she could not have supplied the mammary gland with fat at the expense of fat previously existing in her body; she probably obtained it ultimately from the proteids of her food. (It must be borne in mind that the proteids, carbohydrates, and fats of animal food all contain carbon, hydrogen, and oxygen, but that the proteids differ from the other two classes in also containing nitrogen; an animal could be kept alive on proteids alone, but not on fats and carbohydrates either separately or together,—a dog fed exclusively on pure fat would die of starvation.) More than five-and-twenty years ago, Lawes and Gilbert proved by direct analysis that for every hundred parts of fat in the food of a fattening pig, four hundred and seventy-two parts were stored up as fat during the fattening period, so that it is evident that fat is formed in the body out of something which is not fat. And Liebig had previously proved that the butter present in the milk of a cow was much greater than could be accounted for by the scanty fat present in the grass or other fodder she consumed. There is overwhelming proof that fat is formed anew in the animal body, for two animals fed on the same food will each store up the special kind of fat peculiar to itself. Moreover, dogs fed on foods consisting largely of different fats will exhibit but little variation in the composition of the fat they store up. Subbotin found that a dog fed after a preliminary starvation period, with one thousand grams of spermaceti, of which he absorbed at least eight hundred grams, nevertheless yielded but the merest trace of spermaceti in the fat of his body.

[1] "Physiology."

There is, finally, abundant evidence for concluding that the carbon of the newly-formed fat, equally in the milk of a cow as in the body of an ox, may be supplied (1) from the carbohydrates of the food, or (2) from the carbon surplus of the proteid food, or (3) from fats taken as food which are not the natural constituents of the body-fat.

If the butter fats of milk were derived directly from the food, and transferred without change to the milk, it would be possible to control, through the food, the quality and composition of the butter produced. This was once regarded as possible because such plants as onions and turnips impart, in a short time, characteristic odours to the milk of the cows consuming them; whilst the marked difference in colour and flavour of butter made in the winter when the cows receive only dry food, and in the summer when grazing, has been attributed to differences in the fatty portion of the fodder. On the other hand, variations in the composition of butter ascribed to breed and to individual peculiarities of the cow, are so well defined and so constant, whatever the character of the food, as to render a direct transfer of fats highly improbable. The question of the essential oils of plants is rather beside the point; a peck of onions fed to a cow would scarcely impart so much flavour to the milk as would a piece of onion directly added to it.

One way of elucidating the question is by carefully examining the butter, and noting the changes in its composition which follow radical changes in the food. Crude cotton-seed oil has some characteristic properties by which its presence, when mixed with butter, even in small quantities, may be detected. A cow whose butter had frequently been tested was set apart at the New York Experiment Station, and cotton-seed meal included in her food. The first cotton-seed meal was given on April 18, half a pound, and the quantity gradually increased to 4 lb. on May 1. This amount of meal contained as much fat as the remainder of the ration, and would therefore be likely to contribute an appreciable amount of cotton-seed oil to the butter, if any direct transfer does occur. Inasmuch as the soaps from crude cotton-seed oil have a high viscosity, whilst those from butter fat are usually low in this respect, the test of the viscometer was applied time after time, with the result that, notwithstanding the change in ration, no appreciable change in the viscosity occurred; this was strong evidence that no cotton-seed oil was transferred to the butter. Nor was any cotton-seed oil detected in the butter when subjected to another delicate and reliable test for the presence of the former. In this case, for certain then, there was no transfer of cotton-seed oil from the food to the milk.

The same thing is shown by the constant and uniform difference found in butters from different breeds of cows, and from individual cows in the same herd all receiving the same food and attention. From such data Dr. Babcock concludes that the composition of the butter fats is practically constant for each individual cow under all circumstances of normal feeding. As a matter of fact, quality of butter depends more upon certain physical properties and flavours, which constitute scarcely more than traces of the butter, and which as yet have not even been identified by chemists, than upon the fixed oils which compose it. Nevertheless, these principles, as well as those which determine the colour of butter, are unquestionably influenced by food and treatment. The question remains, How?

When cows are largely fed on watery herbage, brewers' grains, or other food containing a high percentage of water, the milk becomes poorer in solids, or, in other words, the proportion of water in the milk increases. This probably arises from the more watery, or poorer, character of the blood, for water taken into the stomach is absorbed almost immediately by the blood capillaries and lymphatics in the walls of that organ, as is proved by the instant alleviation of thirst when water is drunk. A course of poor watery diet, then, impoverishes the blood, and poor blood leads to the production of poor milk in the mammary gland—of milk, that is, which contains a larger percentage of water than would be the case with better food.

Both the secretion and the excretion of milk are under the control of the nervous system, but the exact mode whereby the nervous influence is exerted remains to be worked out. Indirectly, however, the secretion of milk must be largely affected through the sympathetic nervous system, whose centre is in a chain of nervous elements extending along the general body cavity just beneath the backbone. This system is distinct from, though connected with, the brain and spinal cord, and it largely contributes to, amongst others, the vaso-motor nerves. These nerves are so called because they are connected with the muscular walls of the blood vessels, and, through their influence on the vascular muscles, determine whether the calibre of the vessels shall be increased or diminished; through them, therefore, the quantity of blood which flows along an artery in a given time is regulated. Thus, undue exposure to cold produces an effect on the skin which is conveyed to the nervous centres, the vaso-motor nerves consequently experience a partial paralysis, and are therefore incompetent to control the arterial muscles in one or more organs of the body. The arteries lose their normal tone, more blood passes through them than is compatible with health, that undue distension of the blood vessels known as congestion is set up, and inflammation is the usual result. It may not be the mammary gland, but the lungs or intestines, which are the direct sufferers, nevertheless, the mammary gland is bound to show the effects of such adverse influences in its smaller and poorer secretions. And it is not unlikely that a similar result is more directly produced when undue stimulation of the vaso-motor nerves which control the arteries of the mammary gland leads to an abnormal constriction of these vessels, and so reduces the supply of blood to the secreting cells. In the case of the kidney, indeed, it has been proved, that any irritation of the nerves which control the muscular walls of the blood vessels supplying the organ has the immediate effect of stopping the excretion of urine. But the sympathetic nerves are further of interest in being those through which any unkind treatment of the cow, wilful or otherwise, is bound to show its effect in diminished yields of milk. Ill ventilated, badly drained, or too draughty cow-houses, careless exposure to bad weather, irregular feeding, brutal usage, fast driving, the mad rushing about provoked by the attacks of the ox warble fly, and a variety of other causes, are bound to exert an influence upon the nerves, the effects of which will be unerringly recorded in the milk-pail.

The familiar comparison of a cow to a steam-engine is obviously hardly a fair one. Fuel is supplied to the engine, and work is got out of it; the cow receives food and yields milk; and if the engine be debited with fuel and the cow with food, and the former be credited with work and the latter with milk, this is about as far as the comparison can safely be carried. The intimate and essential relation which is set up between the food of the cow and the structure and composition of the animal is of an order vastly superior to that which exists between the fuel of the engine and the inanimate parts which are set in motion, without undergoing any necessary internal change, as a result of the oxidation of that fuel. In estimating the efficiency of a steam-engine, it is quite sufficient to compare the power theoretically deducible from the complete oxidation of the fuel with that exerted at the driving point on the resistance to be overcome ; the difference between the two, shows the loss due to friction and other causes. But, as Sir J. B. Lawes has pointed out, in estimating the milk-producing capacity of a cow, as shown by comparing the amount of food consumed with the quantity of milk yielded, the live weight of the animal is an important element to be considered, particularly in cases wherein the performances of different cows have to be compared. Were the cow a rigid machine like the steam-engine, and were her milk-producing work effected simply by her mechanically receiving food at one place and discharging milk at another, her weight then would be a matter of secondary and trivial importance. But it is obvious that in the cow not only the absolute live weight, but the percentage increase during lactation, should enter as factors into any problem concerned with estimating and comparing the milk-yielding capacities of different animals.

CHAPTER V.

OF THE MANAGEMENT OF MILK AND CREAM, AND THE MAKING AND PRESERVATION OF BUTTER.

"MILK is an opaque fluid secreted by the mammary glands of the females of animals belonging to the class Mammalia, and adapted to the nourishment of their young. It is of a specific gravity somewhat greater than that of distilled water."

The *average* composition of the milk of the cow may be stated to be:—

Water	87·05
Casein (pure curd)	3·50
Butter	3·70
Milk sugar	4·60
Albumin	0·40
Ash	0·75
	100·00

and in extreme instances it may vary from:—

Water	80·00	to	90·00	per cent.
Casein	3·00	,,	5·00	,,
Butter	1·80	,,	5·20	,,
Milk sugar	3·00	,,	5·50	,,
Albumin	0·30	,,	0·55	,,
Ash	0·70	,,	0·80	,,

"Neither casein nor butter is in solution in milk, but rather in suspension—the butter-fat expressly so. Casein appears to be in the form of an extremely attenuated jelly, owing to lavish absorption of water; but it is not dissolved, or it would pass the membrane of a dialyser. It is soluble in diluted hydrochloric acid, or carbonate of soda, and it is coagulable by rennet, and by lactic acid, and may be precipitated by various acids. Coagulation by rennet, which is the active agent of digestion in the fourth stomach of a calf, is the only form of coagulation that can be employed in cheese-making, for it is the digestive agent alluded to which has so much to do with ripening and mellowing the cheese after it is made.

"Butter-fat, in the form of cream-globules, is easily seen by the aid of a microscope to be in suspension in milk, and each globule is a separate entity. These globules belong to the 'infinitely little' in Nature, for a single pint of milk, containing 4 per cent. of cream, has been estimated to contain no less than the prodigious number of forty thousand millions of them! The diameters of the globules vary a good deal in all milk, and in the milk of different breeds of cows, or in that of different cows of the same breed, sometimes. Sturtevant gives them at $\frac{1}{1440}$ to $\frac{1}{5520}$ of an inch. Milk, indeed, is an emulsion, in which the most valuable ingredient is butter-fat. The specific gravity of milk containing all its cream is about 1·032, whereas that of butter-fat is about ·90, water as a standard being 1·00; and it is this difference in specific gravity which causes cream to rise to the surface of milk that is at rest. Some of the cream-globules, however, have the peculiarity of being stationary, while others appear to gravitate slowly downwards, and hence it is that the whole of them never succeed in reaching the surface of the milk."[1]

Milk from which the supernatant fluid or cream has been removed, is termed skim-milk, and still retains a considerable quantity of coagulable or caseous matter, which may be separated from the serum or whey, by means of a rennet or some acid. This coagulated portion constitutes the curd, and is the basis of cheese. If a rennet be used, and all the portion coagulated by its means be separated, the addition

[1] "The Farm and the Dairy," by J. P. Sheldon, pp. 59, 60.

of vinegar will cause a portion of what was left to coagulate. What remains after both of these coagulated principles have been removed is *whey*, containing sugar of milk, some nitrogenous substance, lactic acid, and various salts.

Some exact information upon the composition of milk is given by Dr. Paul Vieth, Analyst to the Aylesbury Dairy Company, in an elaborate paper published in the "Journal of the Royal Agricultural Society of England," vol. xxv., 1889. He discusses the results of analyses of no less than 84,746 samples of milk, made during a period extending over eight years. The samples analysed were the produce of herds kept upon English dairy farms, and Dr. Vieth records the results in a series of diagrams indicating the total solids, the non-fatty solids, and the fat, according to the monthly averages of eight years. It is premised that the total solids include everything that is valuable in milk; the fat must, in several respects, be considered the most valuable component part; and the percentage amount of non-fatty solids is of particular importance as being the most constant factor, and, therefore, best adapted to serve as a guide to the genuineness and purity of the milk. It is hardly necessary to remind our readers of the fact which may be tersely expressed, so far as milk is concerned, thus :—

$$\text{TOTAL SOLIDS} = \text{NON-FATTY SOLIDS} + \text{FAT.}$$

Dr. Vieth finds, as the result of many analyses, that 12 parts of the non-fatty solids are thus made up on the average :—

Milk sugar	6
Mineral matter, or ash	1
Nitrogenous matter, or proteids	5
	12

Of the nitrogenous matter rather more than two-thirds consists of casein, the specific component part of cheese. From the foregoing figures is calculated the following average percentage composition of the non-fatty solids of milk :—

Milk sugar	50
Mineral matter, or ash	8½
Nitrogenous matter, or proteids	41⅔
	100

The average composition of all the milk received at the Aylesbury Dairy Company's premises at Bayswater indicates a great uniformity in the quantity of non-fatty solids, and a rather marked variation in the percentages of fat, with, of course, a concomitant fluctuation in the amount of total solids. Towards the end of the year—more particularly in November—the milk attains its highest quality, while in the spring months the poorest milk is received, another diminution in quality frequently occurring in July. Of these fluctuations Dr. Vieth offers the following explanation :—In autumn the majority of cows are stale, and they then give a limited quantity of milk, which is of high quality. In spring most of the cows are newly calved, and the fresh grass, forming a very succulent food, causes an increased flow of milk of a poorer description. In summer the cows frequently suffer from either excessive heat, which burns up the pasture, or continuous rain; and these unfavourable conditions are reflected in the production of milk, which becomes inferior both in quantity and quality.

It may be inferred from Dr. Vieth's diagrams, which are constructed on the graphical system, that the average composition of milk, deduced from the results of 84,746 analyses of samples received from 30 to 50 farms during the eight years 1881 to 1888, is:—

$$\begin{array}{lr} \text{Water} & 87\cdot1 \\ \text{Solids} \left\{ \begin{array}{l} \text{Fat} \quad 3\cdot9 \\ \text{Not Fat} \quad 9\cdot0 \end{array} \right\} & 12\cdot9 \\ \hline & 100\cdot0 \end{array}$$

This result is very interesting when placed alongside the general average composition of milk as given in, for example, Professor Sheldon's "Dairy Farming," which is as follows:—

$$\begin{array}{lr} \text{Water} & 87\cdot25 \\ \text{Solids} \left\{ \begin{array}{l} \text{Fat} \quad 3\cdot50 \\ \text{Not Fat} \quad 9\cdot25 \end{array} \right\} & 12\cdot75 \\ \hline & 100\cdot00 \end{array}$$

Further value attaches to Dr. Vieth's paper, in that it separately records the analyses of the morning and evening milk, from a number of distinct farms, for a period of eight years. The particulars are these:—

Farm A, situated in Cheshire; 30 Shorthorns; 2,407 analyses.
Farm B, in Wiltshire; 60 Shorthorns; 2,246 analyses.
Farm C, in Berkshire; 35 Shorthorns; 2,243 analyses.
Farm D, in Wiltshire; 140 Shorthorns; 4,815 analyses.
Farm E, in Wiltshire; 33 Shorthorns and 2 Alderneys; 2,213 analyses.
Farm F, in Berkshire; 57 Shorthorns; 2,063 analyses.
Farm G, in Berkshire; 50 Shorthorns; 2,617 analyses.

It is found that, with exceedingly rare exceptions, the evening milk is richer in every respect than the morning milk. Dr. Vieth is inclined to ascribe this difference to the inequality of the interval between the two milkings; a larger yield of poorer milk being produced after the longer interval from the evening milking to the morning milking. The proportions of morning to evening meal vary from 100 to 94 in the case of farm E, to 100 to 78 in the case of farm A.

A year's analyses of the milk of three distinct breeds of cattle kept at the Aylesbury Dairy Company's farms at Horsham, afforded some valuable comparative results. The herds comprised 84 Shorthorns, of whose milk 1,006 analyses were made; 17 Jerseys, with 236 analyses; and 35 Kerries, with 410 analyses. The yearly average percentage results are summarised in the subjoined tables:—

		Shorthorn.	Jersey.	Kerry.
Morning	Non-fatty solids	9·0	9·6	9·2
	Fat	3·5	5·1	4·0
	Total solids	12·5	14·7	13·2
Evening	Non-fatty solids	9·1	9·6	9·3
	Fat	4·4	6·3	5·1
	Total solids	13·5	15·9	14·4

The points of interest here are the enormous richness of the milk of the Jerseys, and not less the high quality of that of the Kerries, which latter certainly surprised many dairy farmers.

The great uniformity of the specific gravity of milk is very noteworthy. In the case of the mixed yield of a large number of cows it rarely falls outside the limits of 1·030 and 1·034 (water being 1·000). In view of the considerable variations in composition which obtain between, for example, Shorthorn and Jersey milk, this may seem remarkable, but the circumstance is readily explained by the fact that the higher percentage of fat, which tends to lower the specific gravity, is usually associated with an increased amount of non-fatty solids, which exercise an influence in the opposite direction. Though larger variations certainly do occur in the milk of individual cows, yet in the 1,652 samples of Shorthorn, Jersey, and Kerry milk to which reference has been made, only 50 samples registered a specific gravity below 1·030, and only 73 samples—notably of Jersey milk—gave a specific gravity above 1·034. By far the greater number of these exceptional specific gravities were not below 1·029, nor above 1·035; though the lowest actually observed was 1·0240, and the highest was 1·0365.

The quality of milk greatly depends on the quantity which the cows will yield and the nature of the food, but most of all on the breed of the cows. The quantity is, to a certain degree, influenced by the manner in which the cows are milked; and it behoves every dairy-man to pay a little more attention to this important process than he is generally accustomed to do. If a cow is roughly handled, it is not only painful to her, but will also cause her to withhold a portion of her milk; whereas, if it is gently drawn, she will yield it freely. It is of importance that it should be drawn to the last drop, for although we do not pretend to believe what has often been asserted, viz. that the last half-pint is richer in cream than the whole of the rest, we fully admit that whatever milk is left in the udder is liable to coagulate and injure the udder, as well as to lessen the subsequent "meals of milk." It sometimes happens that the cows are restless and fidgety; but they should by no means be harshly or severely treated at such times. If the udder is hard and painful, it should be fomented with lukewarm water, and gently rubbed, by which simple expedient the cow will generally be brought into good temper, and readily yield her milk. It is also proper to feed the cows at the time of milking, for, while eating, they give out their milk with greater freedom. They are also prevented, by the motion of their jaws, from the habit, which some acquire, of withholding their milk, and which, if it be not properly prevented, will soon cause them to become dry.

In this country, it is the general practice to milk cows twice in the course of twenty-four hours throughout the year; but in summer the proper periods may be three in every day, and at intervals as nearly equidistant as possible, viz. very early in the morning, at noon, and a little before the approach of night. It is a well-known fact that cows, when milked thrice in the day, yield more in point of quantity, and milk of as good, if not better quality, than they will give under the common mode of milking them only in the morning and evening. Very particular directions should be given that the cows be driven slowly to the place of milking. If they are hurried in ever so slight a degree, the separation of the milk into its constituent parts will not so readily or perfectly take place. On this account, if the pasture is at a considerable distance, it may, perhaps, be better to milk them in the field than to drive them home. If cleanliness were attended to as much as it ought to be, the udder and teats would be washed with a sponge and water, or preferably dry-rubbed, before the milking commenced.

After the milk is drawn from the cow, it should be carefully strained through a gauze or linen cloth, stretched on an open-bottomed wooden bowl or milk-sieve, into the cream-pans, which should never exceed three inches in depth, although they may be made so wide as to contain any quantity required. The milk-pail should then be rinsed with about a quart of cold water, which also may be poured through the sieve into the milk-dish. If any ill flavour is apprehended from the cows having eaten turnips, &c., the addition of one-eighth part of boiling water to the milk, before it is strained into the dishes, will in a great degree tend to remove it, or the solution of nitre may be used, as already recommended. These pans, when filled, should be set upon the shelves, there to continue until the cream is removed.

"The question of temperature naturally influences the absorbent capacity of milk, for so long as the milk is warmer than the atmosphere of the room, it gives off rather than attracts odours. Cold air coming in contact with warm milk is expanded and rises, and its capacity for holding gases, vapours, and odours is increased, so that it attracts volatile odours from the milk, and may even be made to purify it to some extent, particularly when the milk is stirred about a good deal. And hence it follows that a low atmospheric temperature is the best for the milk-room; but when the temperatures of the milk and of the room are equal, or when that of the milk is the lower, the milk-pans are better covered over to shut off free contact with the air. Milk that has been cooled by water or ice should not be exposed to an atmosphere ten or twenty degrees warmer, for it then becomes a facile condenser and absorbent. While the air is seldom pure enough not to injure milk that is ten degrees colder, it is seldom so impure as to vitiate milk that is ten degrees warmer. It is of course expedient to remove milk at once from the odours of the cow-house to the comparatively pure air of the milk-room, yet there is no special need to do so to avoid contamination, so long as the air is colder than the milk. Yet the odours of the cow-house are often traceable in the milk, and it is thought they get into it by absorption; this may be so, and will be so when the air is warmer than the milk. It is a good thing to have cow-houses well ventilated, and kept as clean as possible."[1]

In the process of milking it should be borne in mind that the milk first drawn from a cow is always thinner, and inferior in quality, to that afterwards obtained, the richness of which increases progressively.

It should also be recollected in the after process, that the portion of cream rising first to the surface is richer in point of quality, and greater in quantity, than that which is yielded in the second equal space of time, and so of the rest; the cream continually decreasing, and becoming thinner and poorer. This is due to the larger globules rising first, the smaller remaining in the milk. If thick milk is diluted with water, it will afford more cream than it would have yielded in its pure state, though its quality will be inferior.

Milk carried about in pails, or other vessels, and thus agitated and partially cooled before it is poured into the milk-pans, never throws up such good and plentiful cream as if it had been put into proper vessels immediately after it came from the cow.

From these fundamental facts several important inferences, some of which have already been hinted at, as serving to direct the proceedings of the dairy, may be deduced.

1. It is evidently of much importance that the cows should be milked as near to the dairy as possible, in order to prevent the necessity of carrying and cooling the milk before it is put into the dishes; and as cows are much hurt by far driving, it must be a great advantage on a dairy-farm, where the practice of house-feeding is not adopted, to have the principal grass fields as near the dairy homesteads as possible.

2. The practice of putting the milk of all the cows of a large dairy into one vessel, as it is milked, there to remain until the whole milking is finished, before any part is put into the milk-pans, is highly injudicious, not only on account of the loss sustained by the agitation and cooling, but also because it prevents the owner of the dairy from distinguishing the good milk from the poor, and guiding him with respect to the profit that he derives from each cow. A better practice, therefore, is to have the milk drawn from each cow put separately, or to have that from only two or three cows put into the creaming-pans as soon as milked, without being mixed with any other.

A small quantity of clear water, cold in summer, and warm in winter, put into the bottom of a milk-pan, will facilitate the rising of the cream; some persons put in a very weak solution of carbonate of soda.

3. If it is intended occasionally, or generally, to make butter of an *extra fine quality*, the milk of all the cows that yield cream of a poor or inferior quality should be rejected, and also the milk that is first drawn from each cow.

[1] "The Farm and the Dairy," Sheldon, pp. 61, 62.

Whenever the making of cheese and of butter of the highest and purest quality is aimed at—and surely this should be the case with every dairyman and with the manager of every factory—it is of the first importance to have the milk absolutely free from all impurities. It has long been known that there is scarcely any substance with which we have to deal which is so liable to be tainted as milk. Hitherto it has been considered that all that was necessary to prevent this tainting was to secure the most perfect cleanliness in everything used in connection with milk, to see to the condition of the cow-houses, their distance from places where bad odours arise, the cleanliness and thorough ventilation of the milk-room, the cleanliness, sweetness, and purity of the vessels used for keeping the milk, of the churns and the cheese-vats. But despite the closest attention to all these points—and they are of such vital importance that attention to them cannot be too much insisted on— recent experience has shown that there are other causes of tainting, which are even perhaps more dangerous in their effects than those named. Such extra taints, if so they may be named, no doubt have been noticed so far as their defects are concerned, but their causes remained unknown; and it is curious to note that these causes have been brought into prominence in consequence of the introduction of new methods of using milk on a large scale, prominent amongst which stands the process for making "Condensed or Preserved Milk," a trade now of great commercial importance. The same gentleman who reported in the Royal Agricultural Society's Journal on the American butter-factories (vol. vii., second series, 1871)—Mr. X. A. Willard of the Cornell University, and of the Maine Agricultural College—in a subsequent report (vol. viii., 1872), on the American milk condensing factories, goes somewhat fully into the tainting of milk.

Mr. Willard points out the general results of the investigations of the learned Hallier and Pasteur, who by means of microscopic observations of the most elaborate character were enabled to show the nature of those causes which are in operation, which change milk from its normal condition, or which render it filthy and unwholesome. These investigations, as may be known to many of our readers, were made in connection with the "germ" theory of disease, with which perhaps more markedly or popularly the name of Pasteur is associated.

From the instant the milk leaves the cow micro-organisms begin their work, and increase with marvellous rapidity. But a very great difference between the condition in which these fungi appear and do their work in milk and in its after products must be here noticed, as it bears closely upon the very point we have now under consideration, namely the tainting of those products. Thus if the spores of the fungi are already in pure milk, to begin with, or are added to the rennet if cheese be made, the fungi appear to do no harm, but in the case of cheese, at all events, seem rather to act in a legitimate way, if it may be so termed; as giving or imparting the peculiar flavour by which the cheese is distinguished. But if the spores come from putrid matter, and gain access to the milk in cheese, they exercise a highly prejudicial influence. Such organisms find in the milk a substance precisely fitted to aid them in farther and rapid development ; which accounts for the amazing readiness with which milk becomes tainted, and the quickness with which decomposition goes on when once it has begun.

We have already adverted to the importance of keeping the bodies of the cows thoroughly clean by currycombing, or by wisping them down with straw ; but this was chiefly in view of the advantage of aiding the functions of the skin—a matter of great importance, which cannot be neglected with impunity. But a new view has arisen since these germ investigations have been made ; for it appears that dirty matter adhering to the skin, and lodging between the folds of the udder, &c., and becoming dry, falls into the milk-pail during the process of milking, and by thus introducing the germs into the milk causes decomposition therein. Not only, therefore, should cows be carefully kept " body clean," by currycombing, so that all adhering matter may be taken off them, but the practice so often permitted of allowing the animals to pass through or stand in mud at gates, edges of water ponds, &c., should not be allowed. This mud is almost invariably mixed with ordure, and the liquid exuviæ of the animals, and, being allowed to remain for perhaps months, becomes tainted in the highest degree. From this will be seen the importance, therefore, not only of having the interior of the cow-byres sweet, and the animals themselves and all the vessels thoroughly clean, but also all the " surroundings " of the dairy,—roads, ponds, &c. The same remark applies to the courts into which the cows are turned for fresh air, when housed for the winter or when kept on the summer soiling system. The dirty practice of wetting the hands with milk in the process of milking should be discouraged.

The taints and flavours of milk and milk-products, and the ripening of the latter, have been shown by recent investigation to be closely allied phenomena. Professor A. Harker, in writing on the subject in the "Journal of the British Dairy Farmers' Association" (1889), remarks that, though till recently regarded as solely a question of chemistry, it is now known that the action of certain living organisms, chiefly minute plants, precedes and is the cause of the elaboration of most of the delicate, and often fugitive compounds, which together make up the so-called flavour of all dairy products. There is no natural product so delicate, so susceptible to minute deteriorating influences, so readily affected by physical changes in itself or its surroundings, so variable from causes apparently beyond our reach, as the material of the dairy farmer—Milk. Of the principal dairy products, cream, condensed milk, butter and cheese, the two latter are of chief importance in a study of flavours, though it is cheese which presents the greatest difficulty and variation in its maturing or "ripening." In even perfectly made butter the action of micro-organisms supervenes after a time and prevents its keeping for long, whilst in badly made butter the action of these organisms is favoured almost as if intentionally, and the unfavourable results are more rapid and pronounced. The action of minute forms of plant life on even the best of butter leads to the liberation of the volatile acids, and produces rancidity, and in the case of the carelessly and imperfectly made article, only partially washed and drained, the sugar and casein left in the butter do but hasten the changes.

The changes which take place in milk are now known to be largely due to the growth within and upon it of specific organisms which have the faculty of multiplying with amazing rapidity. They affect the milk by their living actions, extracting from it those constituents necessary to the building up of their own bodies, and leaving the remaining constituents, which they do not require for that purpose, arranged in new and varied combinations. The organisms are undoubtedly of many different kinds, and unlike each other in their effects upon the milk. From the time the milk is drawn from the udder till the time when coagulation is effected there may be established (by migration from the air or from any source in contact) healthily multiplying colonies of, it may be, some score of different organisms. The primary cause of variability in the product is the presence in the curd and whey of a crowd of living organisms which have been affecting the milk from the first, and are to continue to act upon it after its conversion into curd, and the subsequent matting, draining, pressing, and drying of the cheese. There are two fairly distinct classes of living things which affect the curd and whey of cheese, many of them not found in the milk at all, but the results of an after colonization in the curd itself, either during its crumbling or even after it has been formed in the vat. These two classes are the bacteria on the one hand and the true fungi on the other, and it may be taken generally that the bacteria are present in the milk and curd from the first, and that the fungi come later on, after the former have partially completed their life. Both, however, assist, often simultaneously, in the change of curd into cheese, and subsequently give rise to the flavours which characterize the product.

There is still another pregnant source of evil in connection with the germs which infest the milk and so much deteriorate it. This is the water supply. Now, as a rule on farms, the proper degree of attention is not paid to this ; anything, almost, in the form of water being thought good enough for the stock to drink. On this point the reader will find some remarks in another chapter of this work. Here it is sufficient to say that stagnant water, impure water—even although it be not stagnant but running water,—well water, all of which contain in many instances decaying or decayed organic matters, may, when taken into the system, give rise to products which will greatly injure the quality of the milk.

There has been much speculation as to whether the milk from tuberculous cows can be consumed with impunity, and, having regard to the fact that milk is so frequently employed as food without cooking, more especially for children, there appears to be considerable risk attending its use. Professor Duguid, discussing this subject in " The Journal of the Royal Agricultural Society, 1890," says (page 314) that tuberculous milk must be looked upon as dangerous and likely to be the means of producing the disease (tuberculosis or consumption) in young or weakly subjects consuming it. The chief difficulty in determining whether the milk of any particular cow or cows is dangerous lies in the inability of the veterinary surgeon to say whether there are any tubercular deposits in the udder. Milk may contain these organisms and even a skilled bacteriologist fail to find them; their absence in the few drops which he examines is no guarantee that they may not exist. Recent experiments in the United States have demonstrated that where tuberculous cows showed no signs of the disease in the udder their milk nevertheless proved infective to rabbits and guinea-pigs fed with it. The results of feeding experiments all tend to prove that the milk from tuberculous cows, if given to animals in the uncooked state, possesses a very much higher infective power than the flesh.

Butter.—Milk consists mainly of three component parts, the *butyraceous*, or oily or fatty substance of which butter is composed; *caseous* matter, from which cheese is formed; and the *serum*, or whey. The comparative value of different dairies, and of different cows in each dairy, depends not only on the quantity of milk itself, but also on the quantity of butter or casein it contains. The ingredients named differ materially in specific gravity or weight, and to separate them is the chief object of the dairy. The cream is the lightest, next in specific gravity is the whey, and the curd is the heaviest. The manufacture of butter involves the separation of the butyraceous part, and this is a mere question of gravity. The milk is left undisturbed, and thus the lighter portion mechanically quits the heavier one, and floats on the top. The separation of the curd from the serum—in the manufacture of cheese—involves coagulation followed by precipitation.

The cream, having separated from the other component parts of the milk in about twenty or two-and-twenty hours, in a medium temperature, is carefully skimmed off, and poured into a vessel, until enough is obtained for churning; or the milk alone is let off by taking out a plug in the bottom of the pan. When the cream has been thus collected, it should be placed in a deep, covered vessel, for the action of the air on the surface dries it. It should also be stirred with a stick or spoon, every time a fresh quantity is added. The object of this is to ensure uniformity as to ripening. The time of keeping it depends on the weather. If the cream from each milking has been kept separate, it may remain from two to four days, in warm weather, without being injured; but if sweet cream is mixed with that which is sour, the two ferment and soon become clotted if the churning is delayed beyond three days. This may be in some degree prevented by the stirring; but it is generally considered best to keep separate the cream from each milking, and thus allow each to become ripe of itself. Cream should be churned before it becomes sour, or the delicate flavour of the butter will be injured. When it is on the point of turning a little sour, it is considered "ripe," and then is the time to churn it. The "ripeness" can be tested with litmus paper. Butter from ripened cream has a flavour more matured than that from sweet cream, and the ripened cream churns all the easier.

x

In some counties the separation of the cream from the milk is not thought to be sufficiently complete by this mechanical process, but, after the milk has stood from twelve to twenty-four hours in the pan, it is put over a slow fire, where it remains until it begins to *simmer*, or is about to boil. As soon as the first bubble raises the surface of the cream, the pan is taken off the fire, and put carefully away for eighteen or four-and-twenty hours, in order to cool. At the end of this time, if the quantity of milk is considerable, the cream will be an inch or more in thickness upon the surface. It is then divided with a knife into squares of a convenient size, removed by means of a skimmer, and called *clotted* or *clouted* cream. It is more solid than the cream obtained in the usual way, and has a peculiarly sweet and pleasant taste. The milk thus treated yields one-fourth more cream than is produced in the common way, but this is at the expense of the residue. It more readily churns than cream produced in the usual way, and forms a butter retaining the peculiar taste of the *clouted* cream.

The cream obtained by the ordinary process of setting consists of the butyraceous portion of the milk with some quantity of casein and of the serous fluid, and these must be separated from each other. This has been found to be best effected by agitation. It might be contrived, on a small scale, by means of a bottle, but it is better accomplished by the help of a *churn*. The cream is violently agitated, and the churner works patiently on until some small particles of butter begin to appear, or, in the language of the dairy, the butter *begins to come*.

There is considerable art connected with this apparently simple manipulation. The churning must not be too rapid or violent, nor must it be too slow and gentle. In the first case, and especially in summer, the product would ferment and become ill-tasted; in the latter it would hardly form at all. From forty to forty-five revolutions per minute is about the proper rate of speed in the case of a barrel churn, and this speed should be reduced at both ends of the process of churning. With an "end over end" churn, sixty revolutions per minute are suitable. The temperature should be carefully regarded. In summer it may be 56° F., and in winter 60° to 62°. In summer the churn should be prepared by moistening the inside with cold water; in winter, with warm water. In summer the churning should be done in a cold room; in winter in one whose temperature is about 60°. Miss E. A. Maidment, in her pamphlet, "The Butter Dairy and its Management," says the following table may be safely adopted:—

Temperature of Air (Fahr.).	Temperature of Cream (Fahr.).
66°	55°
64°	56°
62°	57°
60°	58°
58°	59°
55°	60°
50°	61°

All churns should have a valve or a plug through which the evolved gas may be repeatedly allowed to escape; and a small pane of glass, through which, without opening the churn, the state of the cream may be noticed from time to time as the churning proceeds, is also to be recommended.

When the butter begins to form it is seen in small granules, and the pane of glass is no longer clouded but comparatively clear. When the granules have aggregated to the size of mustard-seed, the churn should be brought to rest. This, indeed, is the only stage at which the butter may be thoroughly separated, by washing, from the caseous matter held suspended in the butter-milk. Nearly all the butter-milk should then be let out of the churn through a sieve, clear, cold water should be put in, the churn should be turned a time or two, the water should be let out through the sieve, and the process should be repeated until the water comes out nearly as clear as it went in, by which time the butter will be thoroughly washed. Care, indeed, must be taken not to over-wash it, and so diminish its quantity and make it insipid. The butter may then be taken out in a mass, of clear, golden colour, and in a granular state; and it may be put at once on the butter-worker, in order that the superfluous moisture may be pressed out of it, and the salt worked in. The triple use of a butter-worker, in fact, is to press out the water, press in the salt, and consolidate the butter into a solid, compact mass, without injuring the grain or making the butter greasy. The proportion of salt worked in will vary from one to five per cent.; but when the butter has been perfectly washed, in the manner described, it will keep sweet some time, in a suitable temperature, without any salt at all. Yet is it true that a little salt will improve the taste of butter, as it will that of fresh beef or mutton.

The best authorities consider that butter should not be touched by hand, and indeed few hands are cold enough not to injure the butter more or less. A butter-worker is represented in figs. 64 and 65, page 269.

The following simple rules for butter making are published by the Royal Agricultural Society of England, 13, Hanover Square, London, W.—Prepare churn, butter worker, wooden hands, and sieve as follows:—

(1.) Rinse with cold water. (3.) Rub thoroughly with salt.
(2.) Scald with boiling water. (4.) Rinse with cold water.

Always use a correct thermometer.

The cream, when in the churn, to be at a *temperature* of 56° to 58° F. in summer, and 60° to 62° F. in winter.

The churn should never be more than half full.

Churn at number of revolutions suggested by maker of churn. If none are given, *churn at* 40 *to* 45 *revolutions per minute.* Always churn slowly at first.

Ventilate the churn *freely* and frequently during churning, until no air rushes out when the vent is opened.

x 2

Stop churning immediately the butter comes. This can be ascertained by the sound; if in doubt, *look*.

The butter should now be like grains of mustard seed. Pour in a small quantity of cold water (one pint of water to two quarts of cream) to harden the grains, and give a few more turns to the churn gently.

Draw off the butter-milk, giving plenty of time for draining. Use a straining cloth placed over a hair sieve, so as to prevent any loss, and wash the butter in the churn with plenty of cold water; *then* draw off the water, and repeat the process until the water comes off quite clear.

Make a strong brine (2 to 3 lb. of salt to 1 gallon of water) and pour into the churn through a hair sieve. Rock the churn a few times before drawing off the brine; take the butter out of the churn, put it on the butter worker, and leave it for a few minutes to drain; then work *gently* until all moisture is pressed out.

N.B.—*Never touch the butter with your hands.*

Miss Maidment observes that salting with brine is recommended for ordinary fresh butter as reducing the amount of working necessary to complete the process. "It is dependent for its success on the strength of the brine, the size of the granules of butter, and the length of time they are exposed to its action. The saltness of a butter must be regulated to meet the market in which it finds its customers, but a brine made in the proportion of half a pound of salt to one quart of water will usually, with half an hour's exposure, give saltness enough for ordinary customers. The London trade demands an almost saltless butter, for which four minutes' brining will serve. The quantity of brine made must be sufficient to cover the butter. If by over-churning (though this should not occur) the butter has collected into granules of too large a size, brining is rendered useless, and dry salting must be followed, with $\frac{1}{8}$ to $\frac{1}{4}$ oz. per lb., according to market."

Dr. Anderson recommends the following preparation as not only preventing the butter from becoming tainted and rancid, but also as improving its colour, while it imparts a sweeter or richer taste than could have been effected by the use of common salt only:—

"Let two parts of the best common salt, and of sugar and saltpetre each one part, be completely blended together by beating, and add one ounce of this mixture to every pound of butter. Incorporate it thoroughly in the mass, and close it up for use.

"It will be necessary to keep butter, thus prepared, for two or three weeks before it is used, otherwise it will not taste well; but, if properly cured, according to the above prescription, it will continue so perfectly sweet for three years or more, as not to be distinguished from newly-made salted butter." It is said that in Holland the salt for butter that is intended to be kept is mixed with the milk before it is churned, by which means both its flavour and preservative qualities are more effectually imparted.

Before the butter is put into the firkin it should be made as dry as possible. A thin layer of salt should then be strewed on the bottom of the cask, and each successive layer of the butter thoroughly moulded into that beneath it. When the cask is full, more salt should be strewed over it, and the head put on. If the butter has been previously well freed from the milk, and the salt moulded into it quite dry, it will not shrink in the cask. This is always regarded as one criterion of the goodness of the butter.

The best butter is that which is made during the season of fresh grass; but, with the addition of a certain portion (which experience only can determine) of the juice expressed from the pulp of carrots, or some ground annatto-root (Bixa Orellana), to the cream previously to churning, winter-made butter will acquire the appearance, though not the flavour, of that which has been churned during the prime part of the summer season.

Upon the subject of colouring, Miss Maidment says, "If this *must* be practised, aim at a natural colour and uniformity. The best standard is the natural summer tint, and to maintain this, when the butter would be otherwise pale, add such colour as will make good the deficiency. The best preparation known to the writer is Nicholls's Annatto, of which one drachm will colour from two to eight gallons of cream, according to need. Whatever is used should be carefully estimated and measured in a glass measure (one of two ounces divided in drachms, and costing 9*d*.) and diluted with water (a pint is sufficient), the water to rinse the measure added, and the whole thoroughly mixed with the cream."

The process of making butter by churning *milk and cream together,* which was formerly much practised in Holland, is usually as follows:— The milk is put into deep jars in a cool place, each *meal* or portion milked at one time being kept separate. As soon as there is the least appearance of acidity, the whole is placed in an upright churn. When the butter begins to form in small kernels, the contents of the churn are emptied on a sieve that lets the butter-milk pass through. The butter is then formed into a mass, as before described.

In Ireland the process still is similar in some parts of the country, but the milk is allowed to arrive at a greater degree of acidity. This is a defect.

The practice of making butter from *lappered* (*i.e.*, coagulated) milk is followed in Scotland. The milk is placed in a large barrel and left for from two to three days till a sufficient degree of acidity is attained, and then the *whole* milk is churned. Butter thus made has been very successful in the butter classes at Scottish agricultural shows.

Of the average quantity of butter produced from one cow, or from a dairy of cows, it is impossible to give any accurate estimate. It would vary with the breed, the pasture, and the management. From $2\frac{1}{2}$ to $3\frac{1}{2}$ gallons—10 to 14 quarts—will generally produce about a pound of butter, and a good cow, in order that dairy husbandry may remunerate the farmer, should yield 200 lb. at the least, in the course of the year, this being produced from 600 gallons of milk. A cow, including pasture and hay, can scarcely be fully provided for from less than three acres of tolerably good land, the rent of which, with the taxes, costs, casualties, servants' wages and food, will hardly leave more than a moderate remuneration to the farmer. The reader may be referred, for special information on the subject to Professor Sheldon's work, "The Farm and the Dairy," published by Messrs. G. Bell and Sons.

Some valuable facts concerning yields of butter which have been accumulated through the efforts of the English Jersey Cattle Society (established 1878), may at this stage be noticed. The exhibition of this Society, held on the 15th and 16th of May, 1890, at Kempton Park, Sunbury-on-Thames, is believed to have been the first show in England devoted exclusively to any individual breed of cattle.

The Prize List was drawn up with a view to show—

That the Jersey breed of cattle is well adapted for producing the largest quantity and finest quality of butter.

That when judiciously reared it comes early to maturity, is naturally a small consumer, and will be found most profitable for the dairy.

That English bred animals may be favourably compared with those bred on the Island, where the climate is more genial.

Instead of first, second, and third prizes, three equal premiums of 10*l*. each were given in each class, and in the butter test classes the Society's gold, silver, and bronze medals were awarded in addition to the premiums.

The list of awards contained, in addition to the usual information, the age, live weight, date of birth of last calf, and weight of milk of each animal.

The figures shown by the weighing machine were most instructive and often surprising. As an instance may be noticed the case of a beautiful cow, "Carillon," whose yield of milk in eighteen days would equal her live weight.

Mr. John Frederick Hall, of Sharcombe, Wells, Somersetshire, reported as follows on the butter test:—

The cows competing for the Society's premiums in connection with the butter test arrived in the show ground on Monday, May 12th, and were divided into two classes, viz.:—

Class 11, for cows having had not less than three calves (21 cows).
Class 12, for cows not having had more than two calves (7 cows).

All these cows were milked dry between 6 and 7 o'clock on the Monday evening. Tuesday's milk was drawn and weighed at 8·30 A.M. and 6·30 P.M., and after being raised to a temperature of 80° F., was passed the same evening through Laval's hand-power separator.

Two of these machines completed the process of separation in three hours' working. About a gill of buttermilk was then added to each lot of cream as a "ferment starter."

On Wednesday morning the cream was raised to a uniform temperature of 58°, and at 11 A.M. four churns were put in motion. By 5 o'clock the churning of the twenty-eight samples was finished, and the unsalted butter weighed.

The table of results is shown on page 311, and the table of foods is given on page 312.

BUTTER TEST AT KEMPTON PARK, MAY 15, 1890.—CLASS 11: JERSEY COWS HAVING HAD NOT LESS THAN THREE CALVES.

Catalogue No.	Name of Cow	Exhibitor	Date of Birth	Live weight	Date of last calf	Days in milk	Milk yield	Butter	Butter ratio, viz., lb. milk to 1 lb. butter	Awards
				lb.			lb. oz.	lb. oz.	lb.	
229	Sherry	Dr. H. Watney	Mar. 5, 1883	966	Dec. 8, 1889	156	41 11	2 8¾	16·12	Gold medal and £10
230	Tiny 3rd	S. H. Williams	Mar. 3, 1886	879	April 26, 1890	17	11 11	2 7¼	14·45	Silver medal and £10
218	Bilberry	S. Baxendale	Oct. 27, 1883	1096	Jan. 19, 1890	114	33 12	2 4¼	14·89	Bronze medal and £10
217	Meadow Pride	W. Adams	July 4, 1884	840	April 24, 1890	19	35 6	2 0¾	17·28	
219	Frieda	E. Carter	April 28, 1880	833	Mar. 10, 1890	64	19 6	1 12½	10·87	
220	Stella	E. Carter	Dec. 15, 1883	890	Mar. 11, 1890	63	31 7	2 0¾	15·59	
222	La Presse	Mrs. Crookes	May 27, 1882	863	Feb. 22, 1890	80	24 4	1 6¼	17·24	
223	Mazzini's Pride	Mrs. Crookes	June 12, 1883	892	Mar. 15, 1890	59	32 4	1 11½	8·76	
224	Example 2nd	The Ladies Hope	Feb. 22, 1884	1018	Oct. 11, 1889	214	29 12	1 13	16·41	
225	Sweet Marjorie	John B. Lloyd	Dec. 12, 1883	865	Feb. 22, 1890	80	23 0	1 15¼	11·68	
226	Sunflower	Hugh C. Smith	July 26, 1880	941	June 4, 1889		28 9	1 11¼	16·14	
227	Melissa 2nd	John Swan	April 1, 1886	747	April 4, 1890	39	30 9	1 14½	16·03	
228	Flora	Dr. H. Watney	Mar. 3, 1882	886	April 9, 1890	34	31 5	2 1	15·18	
234	Young Doctress	J. Brutton	June 25, 1883	986	Dec. 7, 1889	157	22 8	1 10½	13·58	
235	Maudie	J. Brutton	July 27, 1880	1028	Mar. 30, 1890	44	35 0	1 12	20·21	
236	M. T. No. 2	J. Brutton	Dec. 1880	998	Dec. 17, 1889	147	19 11	1 9	12·60	
237	Disgusted 3rd	A. C. P. Gurney	April 26, 1885	857	Feb. 27, 1890	75	35 9	1 5½	26·46	
239	Maple	C. and M. Palmer	Feb. 8, 1883	828	April 10, 1890	33	30 1	2 1	14·57	
240	Belle Broughton 3rd	G. W. Palmer	Nov. 6, 1884	888	Feb. 12, 1890	90	20 4	1 8¼	13·36	
241	Mayflower	G. W. Palmer	Jan. 16, 1886	801	Jan. 8, 1890	125	19 4	1 5½	14·32	
242	Wolseley's Fancy	G. W. Palmer	June 15, 1884	808	April 30, 1890	13	33 0	1 10½	19·92	

CLASS 12.—JERSEY COWS HAVING HAD NOT MORE THAN TWO CALVES.

247	Finish	E. Carter	Feb. 9, 1886	857	Jan. 3, 1890	130	33 6	1 15	17·22	Gold medal and £10
244	Blossom 2nd	W. Adams	April 4, 1887	698	April 4, 1890	39	25 13	1 15¼	13·21	Silver medal and £10
254	Cynthia	Mrs. Petkins	Jan. 14, 1887	765	April 20, 1890	28	26 5	1 14½	13·80	Bronze medal and £10
249	Primrose Dame	J. B. Lloyd	Mar. 18, 1887	657	Feb. 13, 1890	89	23 12	1 4½	18·53	
250	Gorse	S. Baxendale	Mar. 24, 1888	882	Mar. 16, 1890	58	23 4	1 13	12·82	
252	Do Good	C. and M. Palmer	Mar. 28, 1887	672	Oct. 2, 1889	223	16 5	0 15½	16·83	
253	Elmhurst Beauty	G. W. Palmer	April 2, 1888	699	April 9, 1890	34	24 8	1 10	15·07	

FOOD RETURN OF JERSEY COWS TESTED FOR BUTTER AT KEMPTON PARK SHOW, MAY, 1890.

Extract from Conditions of Entry:—

"A certificate declaring all the food used for a fortnight before and during the Show must be given to the Society."

No. in Catalogue	Exhibitor, and Particulars of Feeding	Bibby cake	Cotton cake	Decorticated cake	Feeding cake	Linseed cake	Crushed oats	Crushed wheat	Bean meal	Linseed meal	Maize meal	Germ sharps	Bran	Grains	Mangel	Parsnips
217, 244	W. ADAMS. Cattle both grass and stall fed; 7 lb. mixture in addition to other food.	lb. ¼	cwt. ¼*	lb. ..	lb. 2	lb. ..	lb. 3	..	lb. ..	lb. 15*	lb. ..	cwt. ¼*	lb. 6	bush. ..	..	peck ..
234, 235, 236	J. BRUTTON. Cattle out day and night; mangel only given at show.	..	..	..	3	2	2	..	..	..	..	..	2	½	lb. 12	..
219, 220, 247	E. CARTER. At grass, with hay and water.	..	..	4 to 5	..	..	..	..	..	..	..	..	..	..	bush. 1½	..
222, 223	MRS. CROOKES. Cows at pasture, heifers and bulls stall fed.	Cows 4 Heifers 2	..	..	..	..	..	..	..	..	..	..	..	..	..	..
224	THE LADIES HOPE. Cow, stall fed, with a little grass.	..	..	..	..	4½	5	..	..	..	..	..	galls. 2	..	lb. 14 during show.	..
239, 252	C. AND M. PALMER. Cattle, at grass, hay twice daily, no artificial food given for 5 weeks.	..	lb. 3	..	..	..	..	..	..	..	..	..	..	..	..	..
240, 241, 242, 253	G. W. PALMER. Cattle, stall fed until within the last fortnight.	..	..	..	..	..	3	..	..	Cows 4 Heifers 2	3	3	5	..	lb. 10	..
226	HUGH C. SMITH. Cow grass fed.	..	..	..	..	2	peck 1	..	1	..	..	..	bush. ½	..	..	..
227	J. SWAN. Cow stall fed; silage, rye, and tares, 40 lb. per day; during show green food substituted.	..	..	..	..	..	..	..	3	..	..	..	..	..	..	..
228, 229	Dr. H. WATNEY. Cows at grass; Thorley's food also used during show.	..	..	..	..	4 during show. 4	..	..	..	1 during show.	..	lb. 1½	2	..	-⅛	1
218	S. BAXENDALE. Cow at grass.	..	6	3	..	..	..	..	..	..	..	..	..	..	..	..
254	MRS PERKINS. Heifer stall fed, 8 lb. hay, very little grass, and 2 lb. Thorley's cakes.	..	..	..	..	..	7 galls. 2½	..	..	..	..	..	galls. 1½	..	9	..

* For 14 days.

The butter generally was of good colour and quality, but in the case of two cows an excessive use of mangel had destroyed every trace of colour. As regards the quality of the milk, it affords remarkable testimony to the value of the Jersey cow for butter production. The average yield of the twenty-eight animals tested was one pound of butter from a trifle over six quarts of milk.

The butter produce of Jersey Cows.—From the time the English Jersey Cattle Society made its first experiment in butter testing at the Agricultural Hall, Islington, in October, 1886, up to 1897 inclusive, many tests had been made, at which 745 Jersey cows, varying in age from under two to over thirteen years, had been put to proof.

From a tabulated summary of the results of these tests it appears that the development of the butter capacity is very gradual, and it seems probable that even at the end of her fifth year the average Jersey cow has not attained her maximum point of butter production.

The milk yield, which, between the age of two and three years, reaches an average of 2½ gallons, or say 25 lb. per day, continues to increase till the ninth or the tenth year, when it stands at rather over 3½ gallons, or 35 lb. per day. At the same time, the butter shows a corresponding increase from 1 lb. 4 oz. to 1 lb. 14 oz. per day. During the whole of these eight years it appears that the average *richness* of the milk varies but little from a standard of two gallons to the one pound of butter. The average results from the 745 Jersey cows above mentioned are:—

One day's milk . . . 32 lb. 6¼ oz., equal to 3 gallons a day.
One day's butter . . . 1 lb. 11¼ oz.
Butter ratio . . . . 19·02, about 16 pints milk to 1 lb. butter.
Average days in milk, 70.

At or about the age of six years the Jersey appears to attain her prime.—Her milk increases materially in richness, and her yield of butter shows a proportionate advance. It seems probable that she maintains this maximum value for some two or three years afterwards:—

Between 6 and 7 years, average butter ratio, 19·14 lb.
,, 7 and 8 years ,, ,, 18·66 lb.
,, 8 and 9 years ,, ,, 19·12 lb.

This view of the prime age of the butter cow derives further confirmation from a comparison of the ages of those cows which succeeded best throughout the series of ten tests that had been completed up to May, 1890. These are as follows:—

	Years.	Months.	Days.
Average age of 10 first prize cows	7	3	10
,, ,, 10 second prize cows	6	2	0
,, ,, 10 third prize cows	5	8	2
,, ,, 10 reserve number cows	5	5	8

The total number of cows of nine years and upwards is too small to afford a basis for judgment, but it is sufficient to warrant the statement that some Jerseys maintain a high dairy value in their tenth year, or even beyond. The summary table on the next page, extending over twelve years, shows this.

The following cows gave the maximum yields of butter from one day's milk in their respective classes :—

		lb.	oz.
Cows between 2 and 3 years	Mrs. A. P. Norris' Stella	1	15
,, ,, 3 and 4 years	Mr. Simpson's Pandora 11th	2	0½
,, ,, 4 and 5 years	Rev. H. S. Williams' Tiny 3rd	2	7½
,, ,, 5 and 6 years	Mr. Adams' Meadow Pride	2	0¾
,, ,, 6 and 7 years	Mr. Brutton's Baron's Progress	3	5
,, ,, 7 and 8 years	Dr. Watney's Sherry	2	8¾
,, ,, 8 and 9 years	Mr. Carter's Coquette	2	4
,, ,, 9 and 10 years	Mr. Baxendale's Bramble	1	14½
,, ,, 10 and 11 years	Mr. Baxendale's Broom	2	4½
,, ,, 12 and 13 years	Mr. H. C. Smith's Lady Savage	2	0¼

The subjoined table presents a summary of the English Jersey Cattle Society's butter tests from 1886 to 1897 inclusive.

Cows' Ages.	No. tested.	Average days in milk.	Average milk yield.	Average butter yield.	Average butter ratio.
			lb. oz.	lb. oz.	lb. milk to 1 lb. butter.
Over 1 and under 2 years	2	34	15 2	0 13	18·43
,, 2 ,, 3 ,,	43	59	24 15½	1 4½	19·48
,, 3 ,, 4 ,,	87	69	30 4	1 9½	18·87
,, 4 ,, 5 ,,	124	68	31 15¼	1 10¾	19·09
,, 5 ,, 6 ,,	134	72	32 9¼	1 11½	18·93
,, 6 ,, 7 ,,	129	85	33 12½	1 12¼	19·14
,, 7 ,, 8 ,,	96	79	33 6¼	1 12½	18·66
,, 8 ,, 9 ,,	51	72	33 2¾	1 11¾	19·12
,, 9 ,, 10 ,,	31	81	34 0¾	1 13	18·77
,, 10 ,, 11 ,,	27	84	35 10¼	1 14½	18·66
,, 11 ,, 12 ,,	10	82	38 11½	1 14¼	20·46
,, 12 ,, 13 ,,	8	73	35 3¼	1 11¼	20·72
,, 13 ,, 14 ,,	3	54	42 1¼	2 1¾	19·85

Butter is the culmination of the dairyman's art. This great delicacy consists of the natural fats of milk, with some water, and should contain nothing else, except as we choose to flavour it with salt. The perfection of butter-making is to secure these fats, separated from the serum or fluid of the milk, and gathered in a mass, with as little chemical and physical change as possible. Unfortunately, perfection has not yet been reached in this art, and there is always present in butter, mingled with the fats and to some extent dissolved in the water, more or less of the protein or curd, and of the sugar of milk. It is these ingredients which play the mischief with butter, by starting the chemical changes leading to rancidity and decomposition. Whilst, therefore, in nearly all other food-products the presence of protein (because of its high nutritive quality) adds to the value of the article, in the case of butter—if it be placed at all in the list of foods—that which has the highest nutrient value is the poorest in those qualities which go to make fine butter. We buy butter for its fat, and the more fat and the less water and protein, the better it is as butter.

The presence of water in butter is associated with the hardness of the latter. On this point, Mr. F. J. Lloyd says ("Journal of the Bath and West of England Society," 1890—91, page 119), "Hardness depends mainly on the amount of water left in the butter ; this may vary from 9 to 19 per cent. Water being liquid at all ordinary temperatures, the more there is present the softer the butter; in the summer too much is generally left in, which makes the butter soft, while in the winter butter frequently contains far less water, and is too hard. The finer the granules are when brought in the churn, the more water will the butter retain, and no amount of subsequent work on the butter-worker will get rid of it. In summer, therefore, it is necessary to collect the butter into larger granules than in winter, so as to exclude water. And before working it is essential to place the granular butter in a cooling box, not merely to lower the temperature and hence harden the granules, but mainly to allow the excess of water to drain away. In the winter smaller granules and more water are advisable."

The substance, remarks Duclaux, which is taken out of the churn is not pure butter fat, but consists of this material mixed with a little of the serum or whey, small quantities of casein, and milk sugar, as well as phosphate of lime and other mineral salts. In addition, there is water, the quantity of which depends upon the method which has been employed. M. Chevreul's investigations have shown that, in the fresh state, pure butter fat consists only of glycerides, that is, of compounds of glycerin with acids, the latter called fatty acids because they enter into the composition of fats. Some, as stearic, margaric, and oleic acids, have the appearance of fat, whilst others are liquids which dissolve in water. The fatty acids with a fat-like appearance were the first known, because they are relatively easy to isolate and are insoluble in water. Although all kinds of butter, notwithstanding differences in breeds and foods of cattle, contain approximately the same proportions of the same glycerides, yet there are, within very restricted limits, marked variations. The glycerides which have been found in butter are palmitin, olein, stearin, margarin (probably a mixture of palmitin and stearin), caprylin, caprin, caproin, and butyrin. The principal ones are present in about the following percentages:—

Stearin and palmitin	62·8
Olein	27·8
Caprylin and caproin	6·0
Butyrin	3·4
	100·0

Left to itself fresh butter gradually loses its fine and delicate flavour and becomes rancid. What happens chemically is this. In rancid butter, there appears in the free state several strong-smelling acids, one of which, butyric acid, has received its name because its odour is precisely that of rancid butter. Another, caproic acid, is named in reference to the fact that its odour calls to mind that of the goat (Lat. *capra*, a goat). The least traces of these acids in the free state powerfully affect both the flavour and the odour of butter, and it is primarily to their presence that butter owes those unpleasant qualities which are conveyed in the description *rancid*. The circumstances that lead to the development of the acids named are discussed at length by Duclaux, but the subject is too technical to be further followed here. The following results of analyses are recorded by the French investigator:—

PERCENTAGE COMPOSITION OF SAMPLES OF (ISIGNY) BUTTER.

Water	12·40	13·36	12·28
Fat	86·71	85·48	86·76
Milk sugar	0·16	0·20	0·17
Casein and salts	0·73	0·96	0·79
	100·00	100·00	100·00

FAT.

Butyrin	5·90	5·87	5·88
Caproin	3·32	3·40	3·39
Other glycerides	90·78	90·73	90·73
	100·00	100·00	100·00
Free butyric acid (per 1000 parts)	0·093	0·106	0·114

An explanation of the significance of the "iodine number" and of the "viscosity number," as recorded in the table on page 318, is given in the footnote.[1] The table is of interest, as showing that, when

---

[1] Butter, like other natural fats, is a compound of glycerin with palmitic, oleic, stearic, and other acids. When butter, or other natural fat, is boiled with alkali (potash or soda), the fatty acids and the glycerin are separated from each other, the acids combine with the alkali, forming with it a soap (hence the process is termed saponification), and the glycerin is set free. Soaps thus formed from the fat of milk (butter) may be termed "butter soaps," and it is obvious that such soaps may be produced by the action of either potash or soda. Now, aqueous solutions, containing four or five per cent. of potash soaps, become, when rendered slightly alkaline by the addition of caustic potash, quite viscous, and, when the amount of free alkali (caustic potash) is considerable, are completely gelatinised. Each of the fatty acids, stearic, oleic, &c., which enter into the composition of fats and oils, forms with potash a soap whose aqueous solution has a definite degree of viscosity, which may be denoted by a number termed the coefficient of viscosity, or the viscocity number; consequently, the determination of the viscocity of soap solutions furnishes a means of discriminating between different fats and oils, and this is especially true in the case of butter and its substances. A soap made from fifteen grams of stearic or oleic acid, with ten grams of caustic potash, dissolved in water, and diluted up to half a litre (0·88 pint) will form, at a temperature of 68°, a very viscous solution—almost a jelly. Lard, tallow, cotton-seed oil, or olive oil, and all common fats and oils behave similarly, and the same is true of butterine, oleomargarine, and all the commercial substitutes for butter. On the contrary, butter itself when treated in this way gives a limpid solution, the viscosity of which is very slight; but, on account of the variations in the amount of their volatile acids and olein, butters have nevertheless a very wide range, so that small quantities of foreign fats mixed with butter may escape detection. The *volatile* fatty acids of butter (butyric, capric, caproic, and caprylic, but chiefly butyric) tend to reduce the viscocity of its soap solutions, partly on account of their low coefficient of viscosity, and partly on account of their high neutralising power, which leaves less alkali in solution. The determination of the viscocity number is practically effected on the same principle as that of the Nessler test for determining nitrogen in potable waters, namely, a made-up solution is added to till it exactly resembles the solution under trial, and the amount of addition is, of course, known exactly. Hence, the viscocity is determined by means of a solution adjusted to the same viscocity as the one under examination, and the number denoting the viscocity expresses really the number of grams of cane sugar dissolved in water and diluted up to one litre (1·76 pints) to make the test solution.

In the New York Dairy Show trials, samples of the butter from seven pure-bred

subjected to identical tests, the butter from different breeds yields different results.

Holsteins gave an average viscocity of 237, ranging from 112 to 461. Samples of the butter of seven pure-bred Jerseys averaged 74, ranging from 50 to 103. Breed peculiarities are thus clearly indicated. The average viscocity of solutions of soaps from butter fat is about 100; that of lard, tallow, oleomargarine, oil, and of all the commercial substitutes for butter exceeds 1,000; hence, this test affords a valuable means of discrimination. Nevertheless, owing to the great variation in the composition of butter, there is such a wide range in the viscocity of the solutions of butter soaps, that it is possible to adulterate the butter giving the least viscous soap as much as 30 per cent. before the higher limit for butter is passed. But by combining the viscocity test with another test known as Reichert's—which depends upon the amount of fatty acids obtained in a specified manner from 2·5 grams of butter fat—a very trustworthy result is obtained. It may be added, however, that where the viscocity number exceeds 500 there need be no doubt whatever as to adulteration; it is the butter giving viscocities of 200 to 500 that must be regarded with suspicion, which may be allayed or confirmed by applying Reichert's test.

It appears probable that the viscocity of butter soap solutions diminishes with age. Dr. Babcock tested seven samples of butter in May, shortly after churning, and again in October. The following table shows the viscocity of each sample in May, and, immediately below, the viscocity of the same sample in October:—

In May . . . 65, 69, 62, 61, 69, 98, 65; average 65·6.
In October . . 54, 62, 62, 58, 53, 62, 54; average 57·9.

A knowledge of viscocity is of value in indicating changes in the physical constitution of milk, which are often of more importance to the dairyman than are changes in the amount of solids or of fat. Changes in the size of the fat globules or in the viscocity of the milk serum are of this character, but are not indicated by chemical analysis. A low viscocity of the milk serum (that is, the liquid part), associated with large fat globules, favours the economical production of butter. The coefficient of viscocity for the fat of milk increases with the size of the globules, hence the greater the difference between the viscocity of the whole milk and that of the skim-milk the more valuable will the milk be for the production of butter, provided the viscocity of the skim-milk is not very high.

A noteworthy fact is that when warm milk is run through a centrifugal separator, and the skim-milk and cream are caught in the same vessel and thoroughly mixed together again, the product is less viscous than the original milk. This appears to be chiefly due to the breaking up of the fat globules. The extent to which this division takes place is shown by the following determination made by Dr. Babcock of the number of fat globules before and after separation:—

	Viscosity.	Number of globules in ·0001 cubic millimetre.	Per cent. of globules with diameter less than one division of the micrometer.
Milk	267	149	52·7
After separation . . .	248	201	60·3
After second separation .	249	174	70·0

The effect which this division of the fat globules has upon the quantity and quality of the butter made from separated cream is not yet ascertained. Nevertheless, the peculiarity under notice may explain why the same manipulation which produces high-grade butter from ordinary cream often fails when applied to separated cream.

In explanation of the iodine number, it may be premised that the combination of the fatty acids with glycerin are called glycerides, and of these, olein, stearin, palmitin, and butyrin are examples. Now, of the fatty acids found in the glycerides of butter fat, oleic acid is the only one which has the property of absorbing iodine. Each molecule of the acid absorbs, moreover, one molecule of iodine, so that the quantity of iodine absorbed is directly proportional to the quantity of olein present. The variation in this iodine number, particularly in the butter from single cows, is very great, showing the percentage of olein in such butters to range from 27·7 to 52·1. In the New York tests the average for the Holsteins was as high as 46·19, that for the other breeds tested being 35·4 per cent. This high proportion of olein in Holstein butter may account for its softness as compared with Jersey butter. The wide range of the iodine number shows olein to be undoubtedly the

COMPARISONS OF BUTTERS FROM DIFFERENT BREEDS OF COWS.

Breed.	Iodine number.	Melting Point, Degrees Centigrade.	Viscosity number.
Jersey	31·2	34·0	74
Guernsey	31·5	33·3	110
Ayrshire	37·8	33·5	66
Holstein	40·0	33·4	237
All others	35·6	33·8	93
Average of all	35·6	33·7	127

It is obvious that the influence of the breed of the cow upon the composition of the butter fat is no less marked than it is upon the composition of the milk. Moreover, contrary to the general opinion, the quality of the butter does not appear to be materially affected by the character of the food. Among the effects of breed thus noted are those differences in butter which relate to its firmness, resistance to heat, texture or "grain," flavour, and general high quality, by reason of the presence of a larger proportion of the more delicate fats. In all these particulars butter from pure Jersey milk excels, whilst the butter from other breeds follows in the order indicated in the table just given.

It is claimed that facts such as have here been detailed show the great differences which exist in dairy products, the influence of breeds of cattle in causing these differences, and the consequent practical value of a study of this subject when selecting stock for the profitable conduct of any branch of dairying.

CHAPTER VI.

Of the Making and Preservation of Cheese.

THE character of cheese is determined not only by the quality of the milk, but by the skill of the maker, and by the general surroundings. The best season for cheesemaking is the period during which good grass is available, from May to September inclusive. In many large dairies cheese is often manufactured all the year round, but the winter cheeses are generally much inferior in quality to those made during the summer months, although it is probable that good cheese might be made throughout the year, provided the cows were well fed in the winter. Much depends on the ripeness, or mellowness, of the milk, just as in butter-making on the ripeness of the cream. Where cheese is made once a day,—the general rule,—the evening's milk, during warm weather, attains by morning a measure of ripeness

most variable constituent of butter. On the other hand, the amount of stearin in butter appears to have been largely over-estimated, for instead of 30 or 40 per cent., it is probably in most butter less than 3 per cent.

which has a good effect on the character of the cheese. In cold weather this ripening is prevented by the low temperature of the milk during the night, and hence it is, at all events in part, that the cheese of late autumn, winter, and early spring ripens slowly, and is wanting in mellowness. To obviate this, the following course was adopted by a most intelligent cheesemaker, with whom we once had a long conversation on dairying topics, in Canada:—The evening's milk, from October onwards, was warmed to about 84° F., and allowed to stand three or four hours, before being added to the next morning's milk; in this way it acquired the requisite mellowness, which it communicated to the fresh milk of the morning, and the result was that the cheese resembled summer cheese in character and sold for as much money. This question of ripening, indeed, in reference to both milk and cream, in cheese- and butter-making respectively, is one which well merits more study and investigation than it has hitherto received.

When milk has been exposed to the air for a certain time, the duration of which varies, according to the season, it becomes sour and coagulates. The curd which is thus formed may then be either made into butter, by the process of churning, as detailed in the preceding chapter, or it may be merely broken,—when the serum or whey will separate from it,—and, by means of pressure, be converted into cheese. This curd, being composed of both the caseous and the butyraceous matter, constitutes the richest, or what is commonly termed *full-milk* or *whole-milk cheese*. That produced by the curd remaining after the cream has been taken off is necessarily poorer in consequence of the abstraction of the butyraceous substance, and is termed *skim-milk cheese*.

It is known, however, that cheese manufactured from sour milk is hard and ill-flavoured, and means have been devised to curdle it while sweet. With this object various substances have been employed, but the most effectual one hitherto discovered, and consequently the most universally used, is taken from the stomach of calves, and denominated *rennet*. It is the digestive ferment secreted by glands in the internal lining membrane of the fourth stomach of that animal. Even after the animal is dead the glands remain charged with this juice, and, if the stomach is preserved from putrefaction, the fluid retains its coagulating properties for a considerable period. As a matter of fact, the maw or stomach of the calf is preserved by salting after careful cleaning. After the maw has been salted a certain time, it may be taken out and dried, and then it will retain the same property for an indefinite period. A small piece of the maw thus dried is steeped overnight in a few spoonfulls of warm water, and this water will coagulate the milk of four or five cows. Liquid rennet is now prepared of uniform strength, by those who make it a business and a study, and many cheesemakers prefer the prepared article to the crude rennet-skin, a given measure of it accurately coagulating a specified quantity of milk.

Milk coagulates with all acids, but acetic and hydrochloric acids are the most effective. If the dairyman has any reason to doubt the power of his rennet, he may always put it thus to the test. Let him take a portion of the milk, heated to 95° F., and add a small quantity of the water in which the stomach has been soaked; by the quickness of the curdling of the milk, and the form of the flakes produced, he will, after a little experience, form a very accurate judgment of the strength of the rennet, and of the quantity which he must pour into the milk.

The methods of making cheeses in most general use in this country are detailed below, but there are many slight variations in the practice of different dairies even in the same district.

Cheshire Cheese.—The evening's milk is set apart until the following morning, when the cream is skimmed off. The latter is poured into a pan, which has been heated by being placed in the boiling water of a boiler. The new milk obtained early in the morning is poured into the vessel containing the previous evening's milk with the warmed cream, and the temperature of the mixture is brought to about 75° F. Into the vessel is introduced a piece of rennet, which has been kept in warm water since the preceding evening, and in which a little Spanish annatto (a quarter of an ounce is enough for a cheese of sixty pounds) is dissolved. (Marigolds, boiled in milk, are occasionally used for colouring cheese; to which they likewise impart a pleasant flavour. In winter, carrots scraped and boiled in milk, and afterwards strained, will produce a richer colour; but they should be used with moderation, on account of their taste.) The whole is now stirred together, and covered up warm for about an hour, or until it becomes curdled; it is then turned over with a bowl, and broken very small. After standing a little time, the whey is drawn from it, and as soon as the curd becomes somewhat more solid, it is cut into slices and turned over repeatedly, the better to press out the whey.

The curd is then removed from the tub, broken by hand or cut by a curd-breaker into small pieces, and put into a cheese vat, where it is strongly pressed both by hand and with weights, in order to extract the remaining whey. After this it is transferred to another vat, or into the same if it has in the meantime been well scalded, where a similar process of breaking and expressing is repeated, until all the whey is forced from it. The cheese is now turned into a third vat, previously warmed, with a cloth beneath it, and a thin hoop, or binder, put round the upper edge of the cheese, and within the sides of the vat, the cheese itself being previously enclosed in a clean cloth, and its edges placed within the vat, before transfer to the cheese-oven. These various processes occupy about six hours, and eight more are requisite for pressing the cheese, under a weight of 14 or 15 cwt. The cheese during that time should be twice turned in the vat. There are several holes bored in the vat which contains the cheese, and also in the cover of it, through which long skewers pass in every direction, the pressure being still continued. The object of this is to extract every drop of whey. The pressure soon obliterates all these punctures, and the cheese is at length taken from the vat as a firm and solid mass.

On the following morning and evening it must be again turned and pressed; and also on the third day, about the middle of which it should be removed to the salting chamber, where the outside is well rubbed with salt, and a cloth binder passed round it, which is not turned over the upper surface. The cheese is then placed in brine, extending half-way up in a salting-tub, and the upper surface is thickly covered with salt. Here it remains for nearly a week, being turned twice in the day. It is then left to dry for two or three days, during which period it is turned once, being well salted at each turning, and cleaned every day. When taken from the brine, it is put on the salting benches, with a wooden girth round it of nearly the thickness of the cheese, where it stands a few days, during which time it is again salted and turned every day. It is next washed and dried; and, after remaining on the drying benches about seven days, it is once more washed in warm water with a brush, and wiped dry. In a couple of hours after this it is rubbed all over with sweet whey butter, which operation is afterwards frequently repeated; and, lastly, it is deposited in the cheese or store-room,—which should be moderately warm, and sheltered from the access of air, lest the cheese should crack,—and turned every day, until it has become sufficiently hard and firm. These cheeses require to be kept a long time; and, if not forced by artificial means, will scarcely be sufficiently ripe under two or three years.

As a matter of fact there are three different modes of cheese-making followed in Cheshire, known as the *early* ripening, the *medium* ripening, and the *late* ripening processes. There is also a method which produces a cheese that is permeated with "green mould" when ripe, called "Stilton Cheshire;" this, however, is confined to limited districts in the country. The early ripening method is generally followed in the spring of the year, until the middle or end of April; the medium process from that time until late autumn, or until early in June, when the late ripening process is adopted and followed until the end of September, changing again to the medium process as the season advances. The late ripening process is not found to be suitable for spring or late autumn make.

There is a decided difference between these several methods of making. In the early ripening system a larger quantity of rennet is used, more acidity is developed, and less pressure employed than in the other processes. In the medium ripening process a moderate amount of acidity is developed to cause the natural drainage of the whey from the curd when under press. In the late ripening system, on the other hand, the development of acidity is prevented as far as possible, and the whey is got out of the curd by breaking down finer, using more heat, and skewering when under press. In the Stilton Cheshire process a larger quantity of rennet is used and less pressure is employed than in the medium or late ripening systems. The various processes are fully detailed in a pamphlet (price 2*d*.) on Cheshire Cheese Making, by Mr. Joseph Rigby, published by the Royal Agricultural Society of England.

The Dutch make their cheese nearly in the same manner as the Cheshire, excepting that they substitute hydrochloric (or muriatic) acid, which imparts to *Dutch Cheese* that peculiarly sharp and salt flavour by which it has long been characterised. They also leave out the cream.

In making *Gloucester Cheese*, the milk is poured into the proper vessel, immediately after it has been drawn from the cow; but being thought too hot in the summer, it is lowered to a temperature of about 84° or 86° Fahr. by the addition of skim-milk, or sometimes by pouring in water. The rennet is then added at the rate of a pint to 100 gallons of new, or 150 gallons of skim-milk. When the curd is ready, it is cut with the curd-breaker, this being drawn repeatedly through the mass. The whey is then taken out, the curd pressed by hand, and crumbled into small pieces like peas. The curd is next put into vats, which are submitted to the action of the press for ten minutes or a quarter of an hour, until the remaining whey is extracted. The material is then removed into the cheese-tubs, again broken small, and scalded with a pail full of water mixed with whey in the proportion of three parts of water to one of whey, and the whole briskly stirred.

This operation should be performed with great nicety, or the curd is liable to be toughened instead of simply rendered firm. The fluid intended to scald the curd should not be above 96° F., nor should the curd be warmed beyond about 84°. After standing a few minutes for the curd to settle, the liquor is strained off, and the curd collected into a vat; and when the latter is about half filled, a little salt is sprinkled over it, and worked into the cheese. The vat is now filled up, and the whole mass of cheese turned twice or thrice in it, the edges being pared, and the middle rounded at each turning. Lastly, the cheese is put into a cloth, and, after undergoing another pressure, it is carried to the shelves, where it is generally turned once a day, until it becomes sufficiently close and firm to admit of being washed.

In the manufacture of these cheeses, the curd is not so often broken as in the Cheshire—the cheese is not skewered while it is in the press, and part of the cream is usually taken away in order to make butter. The scalding is to wash out any remaining whey, or, perhaps, to dissolve any portion of butter that might have been separated before the rennet had coagulated the milk.

Cheddar Cheese was first made in the village of Cheddar, in the Mendip Hills, in Somersetshire. The process, as now practised, is thus described by Mr. George Gibbons in the "Journal of the Royal Agricultural Society of England," vol. xxv., second series, 1889 :—

As soon as drawn, the milk should be taken to a receiver, about eighteen inches square, placed in the most convenient position outside the dairy, so that by a short open shoot it can pass through the wall into the cheese-tub, being thoroughly strained in the process, and thus doing away with the necessity of the milkers entering the dairy. The evening's milk can generally remain in the cheese-tub during the night; when the temperature is high, an occasional stirring is useful; but in damp, hot, moist weather, or during electrical disturbances, some of it should be placed in other vessels.

In the morning the first duty of the careful cheese-maker will be to examine the condition of the night's milk, and, if acidity be perceptible, the morning's milk only should be heated; as a rule, this is advisable from about the middle of June to the end of August. The night's milk should be skimmed, and the cream put in with the milk which is to be heated in a tin vessel called a warmer, surrounded by hot water in the open boiler, in the boiler house, and in which the whey is also heated. Particular care must be taken not to exceed a temperature of 95°. By this the united milk should be raised to 84°; but by the end of June it may be reduced to 82°. A little sour whey may be added in the earlier and later months, but its regular use cannot be recommended.

When annatto is used, it should be well stirred in, and then sufficient rennet added to coagulate the milk in sixty minutes. The intimate mixing of the milk and rennet is very important, and should usually occupy ten minutes, not only for its thorough incorporation, but also to prevent the cream rising to the surface. It is necessary that the tub should then be covered over till coagulation is complete, in order to guard against a too rapid fall in the temperature of the milk. By the time the curd will break clean over a tubular thermometer, the delicate operation of breaking should begin. This is facilitated by the use of a thin knife, long enough to reach the bottom of the tub, for cutting the curd into squares of about two inches.

This done, it should be left to harden a few minutes and for the whey to separate, when, by the use of a shovel-breaker, the splitting of the curd in its own grain commences. This at first must be done with the greatest caution, or the whey will get white and loss of quality ensue; but the speed should increase as the curd hardens—always taking care that it is regularly broken, and not smashed, until it is the size of a pea, and the whey of a greenish hue; the time of this operation depends somewhat upon the quantity dealt with, but it should take from fifty to sixty minutes. The mass should now be allowed to settle for ten minutes, when with a syphon sufficient whey may be drawn off, as, when heated to not more than 130°, would raise the whole to 90°. During the application of this whey the curd should be well stirred and mixed. A further rest of ten minutes takes place, when enough whey should be drawn off for heating to 130°, and the whey in the tub lowered till it only covers the curd by about two inches. The heated whey should now be poured in a small stream over the curd, the operator taking the utmost care that the whole mass is thoroughly broken up and incorporated with it, the thermometer being frequently used, until it stands at 100°, the limit desired; but the stirring must be continued until the curd becomes shotty and is disposed to sink, the whey showing above it clear and green.

This operation may take from ten to thirty minutes, but should the curd not harden sufficiently fast, and the temperature fall quickly, it would be well to add more hot whey, so as to retain the heat at 100°. The curd may now rest thirty minutes (or, if it is sufficiently acid, a shorter period will do), when all the whey may be let off, and the curd piled as high as possible in the centre of the tub. Carefully wash down all crumbs, strain, and place them on top of the mound. Cover and keep it warm with cloths until it has become sufficiently solid to cut into large pieces which can be turned over without breaking. When this has been done, the whole should be again piled and kept covered for thirty minutes longer, as before; after this it may be removed to the curd-cooler, cut into smaller pieces, and again piled and covered for thirty minutes. This cutting, changing, piling, and covering is continued until the curd presents a rich, dry, mellow, solid appearance, and a perceptible amount of acidity has been developed. This is easily ascertained by taste and smell. It is now ground, and should present a ragged solid curd, dry, but greasy, and if several pieces are pressed together by the hand the fragments should easily fall apart. Fine clean dry salt should be used at the rate of $2\frac{1}{4}$ lb. per 112 lb. of curd, and thoroughly mixed with it.

At this point the temperature of the curd should not be below 70°, and it should be put into the vat or mould, lined with a thin cloth large enough to cover the cheese, placed in the press, where it has a pressure of about 20 cwt., and allowed to remain there until the next morning, when the cloth should be changed, the position of the cheese inverted, and replaced in the press until the following morning. A little fat rubbed over it softens the surface, and is useful in preventing cracks, a square piece of muslin being placed on its top and bottom, and the sides also completely covered with the same material, of sufficient width to draw over the squares $1\frac{1}{2}$ inch, to which it should be neatly sewn. Replace the cheese in the press, where it should continue two days longer. It should then be stoutly bandaged and removed to the warm cheese-room, whence, after being turned daily for six weeks, it should be taken to the cooler room, and turned every other day until three months old, after which, turning once every four or five days is sufficient. Much trouble and damage to the cheese is saved by the use of vats, which open with a key, as made by Brown, of Shepton Mallet, Somerset.

Some successful makers scald at a lower temperature, only raising the first scald to 86° or 88° by whey heated to 120°, stirring the curd to assist the hardening fifteen or twenty minutes. The temperature of the second scald should be 98°, by whey heated to 130°, and it should be stirred until the curd is shotty. It should then be left for twenty minutes, or less, if acidity develops fast. In this case no whey is removed from the curd previous to scalding, except what is required for heating. After the expiration of the time of rest, let all the whey run off; then the usual course is to place the curd in the centre, cutting, turning, covering, and keeping warm, putting it on a rack to drain, placing a board and heavy weights on it to facilitate separation of the whey, promote acidity, and produce a solid curd.

The foregoing descriptions of the manufacture of Cheddar Cheese may be generally followed on small or medium-sized dairy farms; but where large quantities of milk are dealt with, a saving of the heavy laborious work entailed in the lifting and carrying the whey to be heated to and from the boiler is most desirable. As the heating of the milk and whey in the cheese-tub by steam or water is not generally favoured, an improved system, which combines the minimum of labour with the highest results of manufacture, is effected by the use of appliances (fig. 71) made by Mr. E. S. Hindley, of Bourton, Dorset. By this system the quantity of milk or whey required for heating is raised by means of a small centrifugal pump to a tin or copper-tinned vessel called the heater, placed on a level with the top of the tub and partly overhanging it. This has a double bottom, into which steam is introduced. A suitable size for a sixty-cow dairy would be $4\frac{1}{2}$ feet by $2\frac{1}{2}$ feet, and 1 foot deep, thus easily containing 60 gallons. The milk in it can be quickly heated to 95°, which it should never exceed. Then by the opening of a tap it passes into the tub for raising its contents to

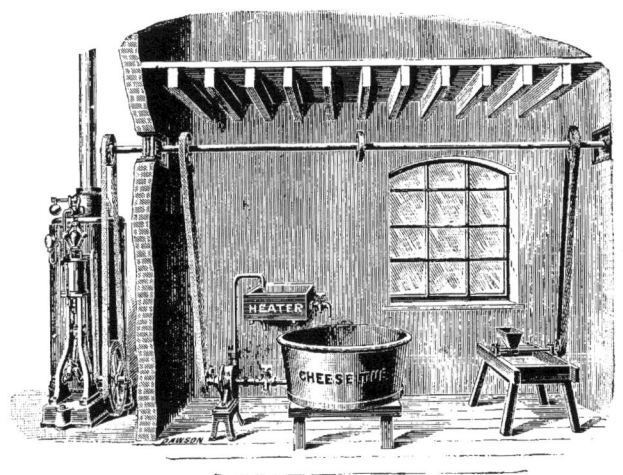

Fig. 71.—Appliances for Making Cheddar Cheese.

the required temperature for renneting, say 84° for the early months, and 82° or 83° later on. The operation of breaking proceeds as before described, but the whey is raised by the same pump into the heater for scalding, and discharged over the curd in the cheese-tub. The lifting and heating of the whey and milk are effected rapidly and without any manual labour, as the pump is worked from a shaft driven by a small steam-engine, the boiler of which supplies the steam to the heater. It also heats all hot water needed, and supplies a jet of steam, which is very useful in the thorough cleansing of utensils. The pump can be cleaned without difficulty, by passing steam and water through it. The shaft also affords a ready means of driving the curd-mill, placed over the curd-cooler; and in those dairies where butter is made, the power is available for driving the separator, churn, and butter-worker. The immense saving of time and trouble, and the certainty with which good results can be obtained by this efficient and comparatively inexpensive system are its chief recommendations.

In a well-managed dairy, where the cheese is properly cured as described, the thin cloths and bandages can be kept on the cheese for transit, as there is thus much less chance of damage; and when the cloth is removed by the retailer or consumer, the cheese should open free from mould, mites, or cracks; and should possess fine mellow texture, sweet aromatic flavour, and pure rich buttery quality, retaining these characteristics if kept for years.

The Canadian system of making Cheddar cheese has, within recent years, attracted considerable attention in this country, and especially has been pursued with marked success in Scotland. The following remarks, contained in a paper on "Cheese-making in Ayrshire," read before the conference of the British Dairy Farmers' Association in 1889, by Mr. R. J. Drummond, are quoted from the "North British Agriculturist":—"In the year 1885 I was engaged as cheese instructor by the Ayrshire Dairy Association, to teach the Canadian system of Cheddar cheese-making. I commenced operations under many difficulties, being a total stranger to both the people and the country: and with this, the quantities of milk were very much less than what I had been in the habit of handling. Instead of having the milk from 500 to 1000 cows, we had to operate with the milk from 25, and not over 60 cows. As a rule I found the people very much prejudiced against anything American; to them it seemed an absurd idea to have to bring an American over to teach the mother country how to improve her cheese.

"The system of cheese-making commonly practised in the county of Ayr at that time was what is commonly known as the Joseph Harding or English Cheddar system, which differs from the Canadian system in many details, and in one particular is essentially different,—viz., the manner in which the necessary acidity in the milk is produced. In the old method a certain quantity of sour whey was added to the milk each day before adding the rennet, and I have no doubt, in my own mind, that this whey was often added when the milk was already acid enough, and the consequence was a spoiled cheese.

"Another objection to this system of adding sour whey was, should the stuff be out of condition one day, the same trouble was inoculated with the milk from day to day, and the result was sure to be great unevenness in the quality of the cheese. The utensils commonly in use were very different to anything I had ever seen before; instead of the oblong cheese vat with double casings, as is used by all the best makers at the present time, a tub, sometimes of tin and sometimes of wood, from 4 to 7 feet in diameter by about 30 inches deep, was universally in use. Instead of being able to heat the milk with warm water or steam, as is commonly done now, a large can of a capacity of from 20 to 30 gallons was filled with cold milk and placed in a common hot-water boiler, and heated sufficiently to bring the whole body of milk in the tub to the desired temperature for adding the rennet. I found that many mistakes were made in the quantity of rennet used, as scarcely any two makers used the same quantity to a given quantity of milk. Instead of having a graduated measure for measuring the rennet, a common tea cup was used for this purpose, and I have found in some dairies as low as 3 oz. of rennet was used to 100 gallons of milk, where in others as high as 6½ oz. was used to the same quantity. This of itself would cause a difference in the quality of the cheese.

"Coagulation and breaking completed, the second heating was effected by dipping the whey from the curd into the can already mentioned, and heated to a temperature of 140° F., and returned to the curd, and thus the process was carried on till the desired temperature was reached. This mode of heating I considered very laborious, and, at the same time, very unsatisfactory, as it is impossible to distribute the heat as evenly through the curd in this way as by heating either with hot water or steam. The other general features of the method do not differ from our own very materially, with the exception that in the old method the curd was allowed to mature in the bottom of the tub, where at the same stage we remove the curd from the vat to what we call a curd-cooler made with a sparred bottom, so as to allow the whey to separate from the curd during the maturing or ripening process. In regard to the quality of cheese on the one method compared with the other, I think that there was some cheese just as fine made in the old way as anything we can possibly make in the new, with one exception, and that is, that the cheese made according to the old method will not toast; instead of the casein melting down with the butter fat, the two become separated, which is very much objected to by the consumer, and, with this, want of uniformity through the whole dairy. This is a very short and imperfect description of how the cheese was made at the time I came into Ayrshire; and I will now give a short description of the system that has been taught by myself for the past four years, and has been the means of bringing this county so prominently to the front as one of the best cheese-making counties in Britain.

"Our duty in this system of cheese-making begins the night before, in having the milk properly set and cooled according to the temperature of the atmosphere, so as to arrive at a given heat the next morning. Our object in this is to secure, at the time we wish to begin work in the morning, that degree of acidity or ripeness essential to the success of the whole operation. We cannot give any definite guide to makers how, or in what quantities, to set their milk, as the whole thing depends on the good judgment of the operator. If he finds that his milk works best at a temperature of 68° F. in the morning, his study the night before should tend toward such a result, and he will soon learn by experience how best to manage the milk in his own individual dairy. I have found in some dairies that the milk worked quite fast enough at a temperature of 64° in the morning, where in others the milk set in the same way would be very much out of condition by being too sweet, causing hours of delay before matured enough to add the rennet. Great care should be taken at this point, making sure that the milk is properly matured before the rennet is added, as impatience at this stage often causes hours of delay in the making of a cheese. I advise taking about six hours from the time the rennet is added till the curd is ready for salting, which means a six hours process; if much longer than this, I have found by experience that it is impossible to obtain the best results. The cream should always be removed from the night's milk in the morning and heated to a temperature of about 84° before returning it to the vat. To do this properly and with safety, the cream should be heated by adding about two-thirds of warm milk as it comes from the cow, to one-third of cream, and passed through the ordinary milk strainers. If colouring matter is used, it should be added fifteen to twenty minutes before the rennet, so as to become thoroughly mingled with the milk before coagulation takes place.

"We use from 4 to $4\frac{1}{2}$ oz. of Hansen's rennet extract to each 100 gallons of milk, at a temperature of 86° in spring and 84° in summer, or enough to coagulate milk firm enough to cut in about forty minutes when in a proper condition. In cutting, great care should be taken not to bruise the curd. I cut lengthwise, then across with perpendicular knife, then with horizontal knife the same way as the perpendicular, leaving the curd in small cubes about the size of ordinary peas. Stirring with the hands should begin immediately after cutting, and continue for ten to fifteen minutes prior to the application of heat. At this stage we use a rake instead of the hands for stirring the curd during the heated process, which lasts about one hour from time of beginning until the desired temperature of 100° or 102° is reached. After heating, the curd should be stirred another twenty minutes, so as to become properly firm before allowing it to settle. We like the curd to lie in the whey fully one hour after allowing it to settle before it is ready for drawing the whey, which is regulated altogether by the condition of the milk at the time the rennet is added. At the first indication of acid, the whey should be removed as quickly as possible. I think at this point lies the greatest secret of cheese-making,—to know when to draw the whey.

"I depend entirely on the hot iron test at this stage, as I consider it the most accurate and reliable guide known to determine when the proper acidity has been developed. To apply this test, take a piece of steel bar about 18 inches long by an inch wide and $\frac{1}{4}$ inch thick, and heat to a black heat; if the iron is too hot, it will burn the curd, if too cold, it will not stick, consequently it is a very simple matter to determine the proper heat. Take a small quantity of the curd from the vat, and compress it tightly in the hand, so as to expel all the whey; press the curd against the iron, and when acid enough, it will draw fine silky threads $\frac{1}{4}$ inch long. At this stage the curd should be removed to the curd-cooler as quickly as possible, and stirred till dry enough to allow it to mat, which generally takes from five to eight minutes. The curd is now allowed to stand in one end of the cooler for thirty minutes, when it is cut into pieces from 6 to 8 inches square, and turned, and so on every half-hour until it is fit for milling. After removing the whey, a new acid makes its appearance in the body of the curd, which seems to depend for its development upon the action of the air, and the presence of which experience has showed to be an essential element in the making of a cheese. This acid should be allowed to develop properly before the addition of salt. To determine when the curd is ready for salting, the hot iron test is again resorted to, and when the curd will draw fine silky threads 1½ inches long, and at the same time have a soft velvety feel when pressed in the hand, the butter fat will not separate with the whey from the curd. I generally advise using 1 lb. of salt to 50 lb. of curd, more or less, according to the condition of the curd. After salting we allow the curd to lie fifteen minutes, so as to allow the salt to be thoroughly dissolved before pressing.

"In the pressing, care should be taken not to press the curd too severely at first, as you are apt to lose some of the butter fat, and with this I do not think that the whey will come away so freely by heavy pressing at first. We advise three days' pressing before cheese is taken to the curing-room. All cheese should have a bath in water at a temperature of 120° next morning after being made, so as to form a good skin to prevent cracking or chipping. The temperature of the curing room should be kept as near 60° as possible at all seasons of the year, and I think it a good plan to ventilate while heating.

"Too much stress cannot be laid on the fact that milk must be pure to obtain satisfactory results. Impurities in milk affect unfavourably not only the value of its products as articles of diet, but the very process which gives the products. At the Dairy Institute near Kilmarnock, we pay the strictest attention to the milking of the cows, to see that each milker washes his hands after the milking of each cow, and, at the same time, the milk is all carefully strained in the byre, and again when delivered in the dairy. The kind of cheese we aim at making is a close cutting, fine flavoured, mild cheese, with good body, and a good cheese in two and a half months, or one that will keep a year."

Reference may here be made to two papers on "Pure Cultures for Cheddar Cheese-Making," by Professor J. R. Campbell, in the "Transactions of the Highland and Agricultural Society of Scotland," 5th series, vol. viii. (1898) and vol. ix. (1899). In the latter volume (p. 219) two important discoveries were announced. The first of these is the fact that it is well within the power of any dairyman to prepare what is practically a pure culture of the same bacterium as that which it had been customary to supply from the laboratory. "The second discovery is that the sour-whey starter, used by some of the successful cheese-makers before the introduction of the American system, was practically a pure culture. These men had, therefore, by empirical methods, attained the same end as that to which we have been led by the more accurate guidance of bacteriological research."

Stilton Cheese has only been introduced since the middle of the eighteenth century. It was first manufactured by a Mrs. Paulet, who resided in the Melton quarter of Leicestershire, and who, being a relative of the landlord of the Bell Inn, at Stilton, on the great North Road, supplied his house with cheese of such singularly superior quality, that it came into demand beyond the consumption of the house, and was sold for as much as half-a-crown a pound.[1] It thus acquired the name of Stilton Cheese; but the mode of making it having been soon discovered, it is now generally manufactured through all the neighbouring counties, and the sale is no longer confined to that place. A great deal of imitation Stilton cheese comes to market, which, though good, is of a very inferior quality to the real Stilton. Its richness depends both on the breed of cows employed, and the quality of the pasture on which they are fed, but especially upon the quantity of *cream* used in the manufacture of it; for, unless a large portion of this is added to the milk, the cheese will be deficient in the essential qualities for which it is celebrated.

For the making of Stilton cheese it is essential that the milk be that of cows fed on good old grass pastures—which ought to have a clay subsoil—supplemented, it may be, by a little cake, say, about 2 lb. per cow per day. It is not advisable to attempt Siltons at all on a very poor pasture, whilst a very rich pasture should be avoided by all but the most expert cheese-makers. A true Stilton, it is well to remember, is not made from unskimmed milk only, but has a certain amount of cream added to it.

The evening's milk is cooled to 65° F. at the time of milking, by means of Laurence's refrigerator, and set until morning in a tin vessel 8 inches deep, 28 inches wide, and 40 inches long, having a hole in the bottom closed with a plug, the stem of which is long enough to stand above the milk when the vessel is full. This must rest on a wood frame 18 inches high, to admit of a bucket being placed underneath in which to draw off the milk.

In the morning draw off one-sixth of the milk through the plughole and put it out of the way (this should not be put in the setting-pan); then draw off a tin bucket full and immerse it in hot water till it is raised to 110° F., stirring occasionally to prevent skimming on the top. Pour it in the setting-pan, draw off another bucketful, and treat in the same way until the vessel is empty. The cream which is left

[1] In "Marshall's Rural Economy of the Midlands," published in 1790, it is stated (2nd edit. vol. i. p. 320):—"Leicestershire is at present celebrated for its cream cheese, known by the name of Stilton cheese. This species of cheese may be said to be a modern produce of the Midland district. Mrs. Paulet, of Wymondham, in the Melton quarter of Leicestershire, the first maker of Stilton cheese, is still [1790] living. Mrs. P., being a relation or acquaintance of the well-known Cooper Thornhill, who formerly kept the Bell at Stilton (in Huntingdonshire, on the great North road from London to Edinburgh), furnished his house with cream cheese; which, being of singularly fine quality, was coveted by his customers; and, through the assistance of Mrs. P., his customers were gratified, at the expense of half-a-crown a pound, with cream cheese of a superior quality, but of what county was not publicly known. Hence it obtained, of course, the name of Stilton cheese. At length, however, the place of produce was discovered, and the art of producing it learnt, by other dairywomen in the neighbourhood. Dalby first took the lead; but it is now made in almost every village in that quarter of Leicestershire, as well as in the neighbouring villages of Rutlandshire. Many tons are made every year. Dalby is said to pay its rent from this produce only. Thus, from a mere circumstance, the produce of an extent of country is changed; and, in this case, very profitably. The sale is no longer confined to Stilton; every innkeeper within fifteen or twenty miles of the district of manufacture is a dealer in Stilton cheese. The price at present is tenpence a pound to the maker, and a shilling to the consumer, who takes it at the maker's weight."

till last must not be raised higher than 98°. Add the whole of the morning's milk after it has been drawn from the cows half an hour, care being taken that the cream does not rise on the new milk—this can be prevented by an occasional slight stirring.

The milk in the setting-pan should be 84° or 86°, according as the outside temperature is high or low at the time the rennet is added. Care must be taken to mix thoroughly, and the pan is covered with a light cloth to prevent the heat escaping. Home-made rennet is usually employed, but Hansen's rennet tablets answer equally as well, though they are more costly.

The milk will be coagulated in from ten to fifteen minutes. In two and a half hours from setting, the curd will be ready to put in the draining-trough; this is done by gently ladling, with a shallow tin bowl holding about half a gallon, the whole contents of the setting-pan into the draining-trough (rods of iron or wood must be placed across the top of the draining-trough to carry the edges of the strainering), in which a piece of *wet* strainering about 48 inches square has been previously placed to receive it. One piece of strainer will hold the curd of about seven gallons of milk.

When the curd is all in the draining-trough, the four corners of the strainer are tied loosely together: the whey remains in the trough half or three-quarters of an hour before drawing off. The curd must then be tied more tightly, which is done by placing the four corners of the strainer together. Very great care must be taken *not to crush* the curd at any time, or the whey will run white, whereas the main object is to keep the whey as green as possible.

The tying will need repeating three or four times during the day, until the curd is sufficiently firm to cut into blocks of the size of half a brick, which will be from six to seven hours from ladling. The pieces must be laid over the bottom of the trough, and in two hours each piece must be carefully turned over, and the whole covered with a light cotton cloth until the following morning. It is now ready to put in the mould (or hoop), but, before doing so, the pieces must be broken to the size of a walnut, mixing salt in the proportion of 8 oz. of salt to 30 lb. of curd. When the hoop is being filled, the curd in the hoop should be occasionally lightly pressed with the hand, and when full it must at once be taken to the draining-room and put in the drainer. Before commencing to fill the hoop it will be necessary to place the latter upon a piece of board, on which to carry it to the drainer; a sinker made of wood, and just sufficiently large to pass easily *inside* the mould, being placed on the top of the curd. As a rule, no other weight should be used, though sometimes it is necessary to do so. No directions as to weights can safely be given, the knowledge must be gained by experience and observation.

After standing three hours, the mould containing the cheese must be turned over on its opposite end, the sinker again being placed on the top. This must be repeated at regular intervals three times a day. At each time of turning on the second and two or three succeeding days the cheese must be skewered through the perforations in the sides of the hoop with a steel skewer about twice the thickness of an ordinary knitting-pin : the outside of the mould containing the cheese must be washed with tepid water, and the drainer thoroughly brushed and washed with hot water every morning.

If the temperature has been kept uniformly at 65°, and the turning and skewering have been properly attended to, the cheese will be ready for the binder about six days from making—here again no precise instructions can be given. The cheese should feel rather elastic under pressure of the fingers ; it will also have left the sides of the mould slightly, so that the latter easily slips off. When the maker is satisfied the cheese is ready for the binder—which is a piece of calico as broad as the cheese is high, and an inch or two longer than will encircle it—the mould containing the cheese is placed on a table, the mould is taken off, and the little holes in the sides of the cheese are filled up by slowly drawing the flat side of a table-knife up and down, applying a slight pressure in doing so, till the side of the cheese is smooth and even.

The binder is now tightly pinned around, and the mould, after being thoroughly cleansed, is again placed over the cheese. The next day this binder must be replaced by a clean one, the side of the cheese being again rubbed over with the knife, and the mould replaced. A clean binder must be put on every day, the mould being discarded after the second day. In very drying weather a light covering must be used for all cheeses in binders.

On the first appearance of coat the knife must no longer be used. In about six or eight days the binder will begin to have dry places upon it, which is a sign the coat is beginning to form. To the eye it will look like little white crinkled patches, but in a few days it will spread all over the cheese, and the coat will then be fully formed. The binder must be used until the coat is perfect.

In very damp thundery weather "slip-cote"—a soft greasy state of the cheese, which will very soon be known by experience—will form instead of the true coat; this must be scraped off with the knife as soon as perceived, and the cheese removed to a cooler place. The best place for the coating process is the setting-dairy, on shelves placed along the wall, except in very hot weather, when a cool moist room is best, with a temperature of about 55° F. The storing-room, which at this time is not fully occupied, is a good place, if care is taken to exclude the midday air.

When the coat is fully formed the cheese must be taken to the drying-room, and placed on deal shelves. It now only requires turning every day, careful attention being paid to cleanliness and draughts. The draught should be rather dry and free, but care must be taken that it is not too free, or cracking of the coat will take place. This latter must be studiously avoided, otherwise the small cheese-fly will deposit its larva in the tiny cracks, and the cheese will be spoilt. The coat should be kept in the same white state as when it came out of the binder. If there is too much moisture in the atmosphere a black mould will form on the coat. This should not be allowed ; more dry air must be admitted, and the cheeses placed farther apart on the shelves. After the cheese has been in the drying-room about twenty days the coat will be firmly fixed, and the cheese must go to the storing-room, to be placed in rows on deal shelves. Here it will only require daily turning, but the shelves must be kept quite clean and free from mites, and careful attention must still be paid to draughts and temperature. In summer it is necessary to exclude the light at mid-day.

Stilton cheeses are sometimes not sufficiently mellowed until they are two years old; and are not accounted to be in good order unless they are decayed, blue, and moist. It is said that small pieces of a mouldy cheese are often inserted into them by means of a *taster*, and that wine or ale is frequently poured over them. Large caulking-pins are also stuck into them to produce the requisite mouldiness. Much of this is bad policy, for the highest perfection is attained when the inside becomes almost as soft as butter and there is not any blue mould save that which develops during the ripening of the cheese.

A Stilton cheese is generally ready for the table in about six months from making. When ready it should have a crinkled light drab coat, it should cut easily with a knife, and if bored it should leave some of the rich soft cheese upon the surface of the borer. It should be well veined with blue, and have a flavour and aroma not to be found in any other cheese of British or foreign make.

For the most recent information the reader is referred to Mr J. Marshall Dugdale's paper on "Stilton Cheese" in the Journal of the Royal Agricultural Society of England, vol. x. (3rd series), 1899.

In making *Wiltshire Cheese*, the milk is used as soon as it is brought from the cow; or, if the temperature is too high, it is lowered by the addition of a little skim-milk. The curd is, in the first place, broken with the hand to various degrees of fineness, according to the kind of cheese intended to be made. For *thin* cheese, it is not reduced so fine as in the Gloucestershire; for the *thick* kind, it is broken still finer; and for *loaves* it is almost crushed to atoms. In the first breaking of the curd, care is taken to let the whey run gradually off, lest it should carry with it more or less of the butter-fat. As the whey rises, it is poured off, and the curd pressed and pared or cut down, three or four times, in slices of about an inch thick, in order that all the whey may be extracted. It is then scalded in the same manner as the Gloucester cheese. In some dairies it is the practice, after the whey is separated, to re-break the curd, and salt it in the liquor; but in others it is taken out of the liquor while warm, and then salted in the vat. The thin sorts are disposed of, with a small handful of salt, in one layer; thick cheeses, with two handfuls, in two layers; and loaves, with the same quantity, in three or four layers, the salt being spread and uniformly rubbed into the curd. In general, Wiltshire cheese is twice salted in the press, beneath which it continues, according to its thickness.

Dunlop cheesemaking, once general in the south-western counties of Scotland, has been almost wholly extinguished by the Cheddar system, which was introduced into Scotland by Harding of Marksbury (*see* page 326), soon after the middle of the nineteenth century, and it may be doubted if now there are any people left who make Dunlop cheese. The Cheddar system, indeed, has taken deep root in Scotland, as in many other countries, and Scotch Cheddars have a high reputation in the English market. The Cheddar system is cosmopolitan in its adaptiveness,—far more so than any other.

First Class *Leicestershire Cheese* cannot be surpassed, either in quality or the price it will command, save perhaps by the best Cheddars and the genuine Stiltons. Low-lying land, having a cold, marly subsoil, carrying a few rushes here and there, and not having been ploughed for a century or two, if ever—land, indeed, whose herbage is ancient and indigenous to the soil—is believed to produce the finest qualities of cheese, whose flavour is rich, clean, full, whose texture is firm and flaky, rather than waxy, and whose mellowness is attractive to the connoisseur in cheese. The methods on which it is made vary a good deal, and each dairymaid follows her own ideas. In spring and autumn the milk is "set" for coagulation at 80° to 84° F., and in summer at 76° to 78°, sufficient rennet being employed to coagulate it in an hour and a quarter or so. Success depends on extracting all the whey, and in curing well without over-salting. The cheese is salted partly in the curd, and partly on the outside of the newly-formed cheese.

Derbyshire Cheese, made on the Carboniferous Limestone soil in the northern half of the county, is a good, practical, sound, every-day article of food, sometimes as excellent and attractive as the Leicestershire cheese. It is made in a similar manner, but the salting is all done on the outside. Mr. George Sheldon, of Low Fields, Derbyshire, made the useful and valuable discovery that the cheese was greatly improved by the addition of about one-fifth of the previous day's curd, which had been allowed to become more or less acid. The discovery was accidental, a few pounds of curd having been mislaid; but the whole year's cheese, upwards of six tons, made in that way, realized 87*s*. per cwt. The milk is coagulated at about 80° F., and in an hour; the curd is then broken, and the whey extracted,—the latter by repeated cutting, "crimming," and pressure—the cheese is in press three or four days, or a week, being turned and dry-clothed once or twice a-day, after which it is taken to the room over the kitchen to ripen. Excellent cheese is made in the Fylde of Lancashire, much on the foregoing plan, the best dairies using sour curd as Mr. Sheldon did in Derbyshire.

Green Cheese is made by steeping in milk two parts of sage with one of marigold leaves and a little parsley, all well bruised; and then mixing it with the curd which is prepared for the press. It may be mixed irregularly or fancifully, according to the pleasure of the maker. The management is in other respects the same as for ordinary cheese. Green cheeses are manufactured in various counties, but only to a small extent.

Skim Cheese was formerly made in the county of Suffolk, whence it was often called *Suffolk Cheese*. The curd is broken in the whey, which is poured off as soon as the former has subsided. The remaining whey, together with the curd, being thrown into a coarse strainer, and exposed for cooling, is then pressed as closely as possible. It is afterwards put into a vat, and again pressed for a few minutes, in order to extract the remaining whey. The curd being thus drained from the whey, is taken out once more, broken as finely as possible, salted, and submitted to the press. The other operations do not materially vary from those adopted in the cheese-making districts; but they are more easily performed on the curd of skim-milk, as it is more readily coagulated and separated from the whey, and requires less subsequent care and pressure than that of milk and cream united. The Suffolk cheese used to form, in general, part of every ship's stores, because it resisted the effects of warm climates better than most others; but it was characterized by "a horny hardness and indigestible quality." A better kind is made in *Dorsetshire*, although the only perceptible difference in management consists in the rennet and the milk being put together cooler; for, by having the milk hot, and immediately applying the rennet, the whey drains so quickly as to impoverish the cheese and render it tough. The old Suffolk cheese was known as "Bang and Thump," and a local rhymester thus described it:—

> " Those that made me were uncivil,
> For they made me harder than the devil.
> Knives won't cut me; fire won't sweat me;
> Dogs bark at me, but cannot eat me."

And the poet Bloomfield said that Suffolk cheese

> " Mocks the weak effort of the bending blade;
> Or in the hog-trough rests in perfect spite,
> Too big to swallow, and too hard to bite."

Whereupon we may conclude that the cheese of the county had a reputation peculiarly its own.

Cream Cheese is generally made in August or September, the milk being at that time richer and better than at other periods. It may, however, be successfully made at almost any time. Cream cheeses are more liable to accident than the poorer sorts, from being chilled or frozen before they become hard, for when the frost once penetrates a cheese of this kind it destroys every good quality, and either makes it ill-tasted or generates putrefaction. Hence, this kind of cheese should always be kept in a warm situation, and particularly preserved from the frost, until it has sweated well, otherwise all the advantage of its rich quality will be completely lost. Cream cheese, however, is in general only wanted for immediate use; and that kind commonly so called is, in fact, nothing else than thick cream drained, and put into a small cheese-vat about an inch and a half in depth, having holes in the bottom to allow any liquid that may exude to pass, and having rushes, or the long grass of Indian corn, so disposed around the cheese, as to admit of its being turned without being broken. It is thus that the celebrated *Bath* and *York* cream cheeses are made, but the greater number of those commonly sold are composed of milk.

Cream cheese, it is obvious, is not "cheese" in the true sense of the term, for it is not coagulated casein, nor is rennet used in its manufacture.

The process of making any kind of cheese, except cream cheese, is much more difficult than that of manufacturing butter, and the quality depends as much, perhaps, on the mode of performing the various operations as on the richness of the milk. The temperature at which the milk is kept before it is formed into cheese, and that at which it is coagulated or turned into curd, are objects of considerable importance in the management of a cheese dairy: the former should not exceed 60° F. nor fall below 50°, and for the latter it should be occasionally from 78° to 82°. If the milk is kept too warm it is apt to become sour, and to give a bad taste to the cheese. If it is allowed to be much colder it becomes difficult to separate the curd from the whey, and the cheese made from it will be soft and insipid. If the curd is coagulated too hot, it becomes tough; much of the butyraceous matter is carried off with the whey, and the cheese is hard and tasteless. The thermometer should, therefore, be employed in every dairy, and, although the servants may at first be prejudiced against it, its evident utility and great simplicity will eventually reconcile them to its use.

The greatest care should be taken to thoroughly extract the whey from the curd, for cheese is apt to heave where any whey remains; and if any part becomes sour, the whole will acquire a disagreeable flavour. Similar effects are produced by the use of an immoderate quantity of rennet. It is also apt to fill the cheese with small vesicles or holes, which imperfection will likewise be produced if the cheese is allowed to remain too long on one side or end.

The cracking of cheese usually arises from the exterior drying too fast, before the interior has become firm. This is commonly caused by the atmosphere of the cheese-room being kept too dry, and at too high a temperature.

Various kinds of *Soft Cheese*,—Camembert, Brie, Neuchâtel, and a score of others which differ more in name than in character,—are made in France and Germany, and some of our ardent reformers have advocated the extensive manufacture of such cheese in England. It is well, however, before rushing into any new practice to reflect whether the public will support it by purchasing the product freely. The French and Germans feed very differently from, and as a rule not half as well as, the English, and the cheeses they prefer are more a relish than a food, far inferior in nutritive value, pound for pound, to the solid, well preserved cheeses of Great Britain. The Continental soft cheeses, too, keep good only a few days, and any falling-off in demand would result in serious loss. The tastes and habits of a nation are not changed in a year, and so far the demand in England for soft cheese of any kind has been, comparatively to that for hard cheese, practically non-existent. The *Slipcote Cheese* of Rutland is, nearer than any other we have, rather like some of the soft cheeses on the other side of the Channel; it has been made during a long period, but so far the demand for it is small. Therefore, it does not appear likely that the soft cheese industry can be held out to our farmers as a tempting pursuit.

Two points of considerable practical interest in the processes of cheese-making are referred to by Mr. George Gibbons in his paper on "Cheddar Cheese Making," both being due to the skill of Dr. F. T. Bond of Gloucester. The first consists in a method of determining the curdling point of milk that has been renneted, with a precision which has not been hitherto attainable. It is founded upon the fact that whilst a drop of ordinary milk, or even of milk that has been renneted up to a certain stage, when allowed to fall gently on the surface of water in a glass vessel, breaks up into rings, and ultimately diffuses completely through the water, yet at a certain point in the development of the curd, which Dr. Bond states has a definite relation to the period when it becomes fit for cutting, the drop falls in a solid mass through the water. This takes place so rapidly as to be distinctly evident within one minute, and Dr. Bond claims that it gives a fixed point by which the effect of rennet can be measured, with a delicacy that is made greater than that of any other method hitherto in use. It is clear that, if this be so, the cheese-maker has at his disposal a simple method by which he can not only measure the strength of any given sample of rennet, but can determine its effects on any given sample or bulk of milk, with much more precision than hitherto, since the tests usually employed for estimating the fitness of the curd for cutting, though sufficient for this purpose, are not exact enough to allow of their being used to measure the strength of rennet with any delicacy.

The practical cheese-maker is recommended to study two papers by Dr. Bond—the one on "Acidity in Milk," the other on "The Work of Acidity in Cheese-making"—in the "Journal of the Royal Agricultural Society," vol. ii., 3rd series, 1891; also his paper on "Germs in the Dairy" in the same Journal, vol. vii., 1896.

The second contribution referred to, is a method of measuring the acidity of milk, whey, &c., by the use of a standard solution of an alkali combined with a colouring agent called an *indicator*. This method, which is familiar in the chemical laboratory, has been adopted by Dr. Bond for use in the dairy, in a way which makes it easily workable by any intelligent person who will take a little trouble to master its details. The control of acidity in cheese-making is a problem which has puzzled many a maker, and if further investigation should confirm the trustworthiness of this method and its applicability to general use, there can be no doubt that the claim which Dr. Bond makes to have added by these two contributions to the precision with which cheese-making can be conducted, will be fully substantiated.

A new kind of cheese, called "Oleomargarine Cheese," has been recently invented in American dairies,—the object being to economize the skim-milk by using lard and other fat of animals to replace the butter taken away in the skimming. The following description we borrow from the pages of "The Farmer":—

"This cheese is made from milk which has been set for cream and skimmed, the cream being turned into butter, and oleomargarine added to replace the material fat of the milk which has been taken off for butter making. The object of adding oleomargarine is to so improve the skim-milk that a cheese may be made from it which is mellow and palatable, and which will resemble in texture and meatiness a whole-milk cheese. Oleomargarine can be bought at about 14 cents per pound, and as a much less quantity is added to the skim-milk than the original butter taken off, the difference in price as well as quantity of the two articles constitutes the profit to be derived from the management of milk on this system as compared with whole-milk cheese-making.

"But, first, how is oleomargarine made? A gentleman who has recently paid a visit to an oleomargarine factory in Hartford, Connecticut, writing in an American contemporary, says:—'Only the very best fresh beef suet from the caul and kidneys is used at this establishment, and of this there is bought and used daily about 600 lb., which is procured fresh every morning from the slaughterer. This fat is first mechanically cut up by means of a machine, which rapidly reduces it to a pulp, so finely ground that it resembles thick cream in consistency. It is then placed in open tanks of sheet iron holding 700 lb. each, which are heated by steam. This thoroughly dissolves out all the oily matter from the cellular tissue, the fatty matter floating on top being drawn or skimmed off carefully and allowed to cool in large vessels. By slow cooling the fat crystallizes, and the more solid margarin and stearin are separated from the olein, which remains diffused through the mass. The semi-solid mass is then put into strong bags of new cotton cloth, the variety known as "Pequot A" being used for this purpose. These bags, which hold about two pounds each, are then placed in a powerful press, which separates the lighter oil, forming the essential principle of butter, from the stearin, which is the harder and heavier product. As it flows from the press the oil is clear, yellow in colour, tasteless, and without odour, having been so refined in the several manipulations that all smell or taste of suet is entirely removed from the oil. The residue in the bags, which is stearin, is sold to be used for hardening lard sent to the Southern market and warm climates.'

"This yellow, tasteless, odourless oil is what is used in the skim-milk for making the so-called oleomargarine cheese, and when prepared as above stated, there is no reason why it is not as clean and as wholesome as the butter fat which comes from the udder of the cow. The most scrupulous neatness is observed in its production, the greatest care being taken to use only the freshest and best suet to be obtained from healthy fat animals slaughtered for beef. Indeed, old, tainted, refuse grease cannot be successfully employed, and if such were used the oleomargarine business would soon come to an end.

"The oleomargarine cheese is said to be a good-flavoured meaty article, having remarkable keeping qualities, and retaining its flavour much longer than the whole-milk cheese. The method of improving skim-milk by the use of oleomargarine is so effective that it is believed a considerable quantity of the new kind of cheese will be constantly thrown on the American markets from the creameries."

In commenting upon it, "Moore's Rural New Yorker" says:—

"From all we have been able to learn concerning oleomargarine, we have seen nothing as yet that would seem to prove that it is unclean or unwholesome; and as a further proof of our confidence that this is so, we may add that for several months past we have been using freely oleomargarine cheese on our table, and find it not only very palatable but wholesome. We believe that all foods, however, should be sold under their proper name, and so of oleomargarine cheese; and while there may be nothing in the flavour or texture to distinguish it from other cheese, it is just and proper that the consumer should know what he is buying, and thus, if he have prejudices against any particular kind of food, he may have full liberty to avoid it."

It is none the less true that the reputation of American cheese in England has suffered most seriously on account of this imitation article, and that the loss inflicted on American dairy farmers in this way is incalculable. It has been rumoured that oleomargarine cheese has been made in a factory in one of the southern counties of England. If the rumour be true we may venture to hope that the practice will be discontinued, or else English cheese will suffer, as American has done, in that popular form of reputation which makes the sale.

From a report on the dairying industry of the United States, which was circulated at the Paris Exhibition, 1889, we take the subjoined interesting details:—As regards external trade, statistics show that the quantity of butter exported is undergoing a constant and rapid decrease, amounting for the entire Union to as much as 62 per cent. in the six years, 1883 to 1888. This shows that the States are depending less and less upon foreign demand. Another significant fact is the almost complete suppression of oleomargarine. The trade in this substance was enormous at the time of the passing of the law against it in November, 1886, and it is largely due to the efforts of a commission appointed to ensure strict compliance with the law that the oleomargarine industry has been almost ruined in so short a time. Seven-eighths of all the dairy produce exported by the United States is from the State of New York. The total value of the dairy produce of the entire Union for 1888 was estimated at about 76,000,000$l.$ sterling, whilst the total value of the dairy products sold upon the New York market during the same year was only $8\frac{1}{2}$ millions sterling, or a ninth of the total produce. The development of the cheese industry of the United States dates virtually from the first establishment of butter factories rather less than thirty years ago. Most of those now existing are worked upon the co-operative system. The great bulk of the cheese is of the kind known as American Cheddar. This make is cylindrical, flat, from 45 lb. to 90 lb. in weight, about 18 or 19 inches in diameter, and 6 inches deep. There is also the Young America, which has the same shape as the Cheddar; but is so much smaller that it is possible to pack five Young Americas in a box which would only hold one American Cheddar. Fancy cheese is not made, save on a very limited scale; but Limburger, Stilton, Edam, Pineapple, Neufchâtel, Swiss, or cream cheeses are variously produced in different localities. The American ideal of a good dairy cow is an animal of

1,000 lb. live weight, producing 30 per cent. of her weight in butter, or half her entire weight in cheese, in a year. Jerseys, Guernseys, Ayrshires, and Holsteins are largely in favour.

The following details are taken from an article by Major Henry E. Alvord on differences in milk products (cheese and butter), which appeared in the report of the Connecticut Board of Agriculture for 1888 :—Premising that good cheese is made from whole milk, or that from which no part of the cream has been taken, and that in old times little else was thought of, it is pointed out that so many inferior kinds are now made that the term " full cream cheese " is given to the standard product of the first quality. The differences now to be discussed are not those incident to the processes which result in " skims" and " filled " cheese (lard or oil substituted for fat removed in cream), but relate to the variations occurring in the quantity and quality of full cream cheese made from an equal weight of whole milk from different breeds of cows. Inasmuch as in well-made cheese a very large proportion of the total solids of the milk is secured in the product—nearly all the casein and fat, though most of the sugar escapes in the whey—it follows that the milk which is richest in total solids will make the most cheese per cwt. of milk, and the general statement is true that milk best suited to butter is most profitable for cheese. The data regarding cheese made from the milk of pure-bred cows of different breeds are meagre, but the principle stated is borne out by experience with Jersey milk. The general average in good cheese-making districts is 10 lb. of cheese to every cwt. of milk; with milk from pure Jerseys, in large number, on the common factory plan, it has been found that the same weight of milk will give over 12 lb. of cheese. At several recent shows in Canada milk from selected cows of different breeds has been tested with regard to its available curd, or cheese-making qualities, and although the animals have been few in number, an enumeration of the general results is not without interest. The order of merit as cheese-makers came as follow :—First trial—Jerseys, Shorthorns, Ayrshires, Guernseys, Devons, Galloways, Holsteins, Polled Aberdeens. Second trial—Jerseys, Ayrshires, Shorthorns, Holsteins. Third trial—Jerseys, Ayrshires, Devons. In the second trial the Ayrshires led in the quantity of curd without fat, but with curd and fat together took second place.

Little information, indeed, is obtainable as to the merits of different breeds of cattle as regards the quantity of cheese made from their milk, although it appears to be a fact that cheese made from the milk of Jersey cows is so much richer in both casein (proteids) and fat, that it is worth a cent a pound more than the average full-cream cheese of America, as an article of nutritious food. Upon this point it is interesting to recall the words of the late Professor L. B. Arnold :— " The business of the Jersey cow is emphatically that of butter-making. Her milk, however, is rich in cheese-matter, and, contrary to the general belief, is capable of making as fine cheese as it does butter. It is a new feature, worthy of note in the uses of this breed of cattle, that their milk can, without the waste of its buttery matter, be converted into a strictly fancy cheese, as rich as English Stilton. Analyses of cheese from pure Jersey milk have shown over 40 per cent. of fat." The table here given—taken from the Connecticut report—affords information as to the extent to which different kinds of cheese may vary in composition.

TABLE OF ANALYSES OF DIFFERENT KINDS OF CHEESE.

Description of Cheese, 100 lb.	Water, lb.	Fat, lb.	Protein, or Curd, lb.	Ash, lb.
1. Average of 83 samples, full cream cheese.	35·75	30·43	27·16	4·13
2. Average of 21 samples, New York State Dairy Commissioner's Report	27·82	28·61	38·10	4·39
3. Full cream (Flint's Dairy Farming, of pure Jersey milk)	38·46	31·86	25·87	8·81
4. Full cream, Premium at New York State Fair	28·37	31·28	36·52	3·83
5. Full cream, Premium at New York State Fair	28·62	29·90	37·66	3·82
6. Full cream, Premium at New York State Fair	33·75	28·95	33·70	3·60
7. Full cream	28·11	41·03	28·18	2·68
8. English average, Sir L. Playfair	38·78	25·30	31·02	4·90
9. English Cheddar, 2 years old, Prof. Johnston	36·04	30·40	28·98	4·58
10. English Double Gloucester, 1 year old, Sir L. Playfair	35·81	21·97	37·96	4·25
11. English North Wilts, 1 year old, Prof. Johnston	36·34	28·09	31·12	4·41
12. Half-skim, average of 8 English samples	46·82	20·54	27·62	3·05
13. Half-skim, New York State	38·35	19·93	38·48	3·25
14. Skim-milk, average of 9 English samples	48·02	8·41	32·65	4·12
15. Skim-milk, English, 1 year old	43·82	5·98	45·04	5·18
16. Whey cheese, average of 6 samples.	23·57	16·26	8·88	4·76

The subjoined analyses are given by Duclaux:—

ANALYSES OF CONTINENTAL CHEESE.

	Grana.	Dutch.	Gruyère.	Old Cantal.	Brie.
Water	32·56	35·37	36·00	36·26	53·95
Fat	21·75	24·72	29·29	31·70	24·60
Casein, insoluble	22·12	25·69	26·51	23·18	12·44
,, soluble (see p. 278)	18·50	8·43	4·33	1·41	4·85
Chloride of sodium	1·65	2·89	0·57	2·23	3·26
Other salts	3·42	2·90	3·30	2·22	0·90
	100·00	100·00	100·00	100·00	100·00

Ripening of Cheese.—Cheddar cheese will ripen in any time from three months to a year or more, according to the conditions of its manufacture. The rule is that a cheese which takes long to ripen will, if well made, keep a long time; a quicker-ripening cheese will not keep good so long. A firm of Scottish cheese-makers, writing to us on this subject, say, "We always, when we wish cheese quickly into the market, make it rather acid. The sweet-made cheese takes longer to ripen, and is often as of fine a flavour when matured." A fine Cheddar, however, will keep well for a couple of years, though, perhaps, it was ripe in three months. What we want everywhere, indeed, is cheese early to ripen and slow to decay,—cheese that can be turned into money before it has lost more than 10 per cent. of its weight in drying, or that will keep well, if necessary, till a market is found for it. A temperature of 70° F., there or thereabout, has been found to ripen Cheddar cheese admirably, but it seldom happens that the atmosphere of the cheese-room is strictly maintained at that point, nor indeed is it necessary that it should be.

It is a good practice to place thin white paper upon the shelves on which the cheeses are laid, because when new they sometimes adhere to the board, and communicate a dampness to it that is prejudicial. The paper also promotes drying. At a more advanced stage the cheese may be laid upon straw, but at first this would sink into and deface the surface.

In a paper upon "The Ripening of Cheese: its Nature and Control," by Dr. Bernard Dyer, ("Journal of the Bath and West of England Society, 1891,") it is pointed out that taints in butter are generally produced by changes brought about by microscopically minute organisms or ferments, and that the main point to be kept in view in butter-making is to render as powerless or inert as possible the organisms with which we cannot keep milk and butter from becoming contaminated. Dr. Dyer proceeds :—

"In cheese-making the case is altogether different. There we are altogether dependent upon organisms, for the ripening of cheese is effected wholly by organisms or living ferments of various kinds—some of them bacteria, some of them moulds. The flavour and texture of cheese are determined by the particular organisms present, and the facilities afforded by one race or another to become dominant. It is the variety of the organisms, and the variation of the conditions in which they are placed, that account for the many kinds of cheese that can be made from one and the same raw material, milk.

"Not unfrequently we hear the special quality of the cheese of a given district attributed to special excellence in the local pasture. This notion (in the opinion of many) is not well founded. It is far more likely to be due in most cases to the special local prevalence of certain varieties of organized ferments. This view is supported by sound facts. It was formerly held that many foreign fancy cheeses could only be made in their native homes, attempts to imitate them elsewhere, even when the various steps of the process were carefully followed, having failed ; the characteristic ripenings required did not take place—*and the characteristic moulds were not developed*. It has, however, been found that if the moulds are transplanted from one dairy to another, and the proper conditions are then observed, the characteristic ripening required *does* take place, simultaneously with the growth of the characteristic moulds. Every one knows the difficulty of starting Stilton cheese-making in a dairy in which it has not been made before, and which is not in a Stilton district. The process of manipulation may be properly learned, and faithfully imitated, and yet the desired flavour is not obtained. The cheese may be rich and good, but it is not 'real Stilton.' Instances, however, can be now pointed to in which after previous failure the would-be Stilton-maker has succeeded by adopting the scientific process of inoculating his curd with fragments of well-ripened Stilton cheese of first rate quality from a good Stilton dairy, with the ultimate result that the necessary organisms have been acclimatised in their new home, and the making of Stilton has become a pronounced success. The very atmosphere of an old dairy is probably thronged with the germs of the organisms, that have long been at work, hourly and daily, and yearly, in its ripening-room. Many of these organisms have been individually studied by biologists abroad, though little scientific work has been directed to them here, and the sum total of the knowledge yet gained about them is very small. But the mere discovery of the principle that specific ripening is caused by specific organisms working under specific conditions, apart from a knowledge of individual species themselves, has already produced such practical results as these.

"The recognition of the fact that we depend on micro-organisms for the conversion of curd into cheese, and of the further fact that the culture of these organisms is greatly under our own control, at once imparts to the empirical processes of cheese-making an interest which to the intelligent dairy farmer they never previously possessed. It throws a new light upon his operations, and awakens observations and a desire for experiment in a far more systematic fashion than was formerly possible, and it is probable that in another half century practical cheese-making will be as much controlled by scientific principles as is already the case in brewing, a process in which the study of micro-organisms and their effect has assumed a great regulating influence.

"To illustrate how close is the correspondence between science and practice in some methods of cheese-making, we may briefly compare two different methods practised in Cheshire, viz., the production of old-fashioned late-ripening cheese, and that of the more modern early-ripening cheese. We will then see how the results arrived at in practice correspond with those which we should theoretically expect to happen from what is known of the general properties of organized ferments and of their work in producing organic changes more or less rapid—for the ripening of cheese is, as already said, essentially a series of organic changes wrought by living organisms. In a dairy in which old-fashioned late-ripening cheese is made, the greatest care is taken to strain and rapidly cool the evening milk, to keep it cool all night, and even to keep it covered till morning. The rennet is so proportioned as to produce a curd that will separate cleanly and firmly from the whey, very little acidity is allowed to develop, and the greatest possible care is taken to remove as thoroughly as possible the whey from the curd.

"In making early-ripening cheese—we are considering, for contrast's sake, the case of very early cheese, such as ripens off in a few weeks (and it may be added rots if it is kept long after it is "ripe"), the milk of the evening is not cooled, and no special care is taken to shield it during the night from the air. By no means is the same jealous care taken in getting a clean firm curd, as is taken with late-ripening cheese; acidity is allowed to develop freely, and the separation of the whey is not effected with anything like the same perfection—indeed much that might be removed is allowed to remain.

"What should we, scientifically speaking, expect to follow, in each case? Milk, before it reaches the cheese tub, is already largely infected by organisms from the dust of the air, like everything else that is exposed. If it is cooled down at once and kept cool, these organisms are kept in check, and increase and multiply but slowly. But if the milk be left with its natural heat in it, to cool down spontaneously, its average temperature during the night will be far higher, and such as to favour the growth and multiplication of the organisms in it—the rate of whose increase under such conditions is enormously great. Furthermore, if it is at the same time freely exposed to the dust of the air, it is all night receiving a fresh access of germs. The consequence is that when the milk is "set" in the morning, the curd thrown down in the one case contains but few organisms, while in the other it contains probably many thousand times more. Each of these organisms that is entangled in the curd may be regarded as a starting-point for change or fermentation. Consequently it is easy to understand that this one difference of cooling or not cooling the night's milk may in itself be expected to enormously influence the rapidity of the 'ripening' process. Then, again, curd freed from whey is far less fermentable than curd containing it. It is the milk-sugar of the whey that in itself mainly nourishes many milk ferments, such as the lactic ferment, and probably many more. By leaving the whey longer in contact with the curd and allowing it to sour, we breed a greater crop of these organisms than if we run it off quickly. And if, in addition, we drain and press the curd but imperfectly, we not only introduce into the cheese more organisms, but we give them more pabulum to feed on. In the one case we have a cheese composed as nearly as we can make it of curd and fat only with but little whey; in the other we have more whey left in, and consequently more milk-sugar and soluble saline matters,—so that the whole mass to be ripened is not the same. We should not expect, therefore, the ripening process to be the same, apart from mere speed. And it is not. The fine flavour of old cheese is never developed in the extremely fast-ripening cheese we are now speaking of. Before the special and more delicate organisms that would give this flavour have time to do their work, the ground is occupied by the, so to speak, ranker and coarser organisms that by strength of numbers and of congenial food at once assume the mastery. A sort of parallel is found in the grass-field. If we sow on a good old pasture coarse rank-growing grasses they soon crowd out the finer and less vigorous grasses and overrun the field. Even if we do not sow rank grasses, but merely manure an old mixed pasture with heavy dressings of rich manure, like nitrate of soda, we know that the coarser and stronger grasses will be so encouraged as to gradually crowd out the clovers and finer grasses. So, doubtless, it is with cheese—reading organisms for grasses, and whey and its constituents for manure.

"When cheeses of intermediate ripening speed are made, it is by observing (or neglecting) such precautions as have been mentioned in an intermediate degree.

"The hastening of cheese-ripening within moderate limits—doubtless a desirable thing in the eyes of many farmers—was arrived at gradually, but of course empirically, and as the result more or less of accident followed by shrewd observation. Had the nature of cheese-ripening, however, been earlier understood, the manufacture of early ripening cheese might have been devised long before. As it is, the knowledge now possessed, limited though it is, enables the control of cheese-making to be effected far more intelligently than was possible before biology began to throw light upon the many interesting phenomena involved in it."

CHAPTER VII.

On the Produce of a Dairy.

THE produce of a dairy is to be regarded in a twofold view, as concerning *quantity* and *value*. Both depend in a great degree upon management; for if the cow is injudiciously treated, or the butter and cheese are badly made, both the product and the price will be materially diminished. There is no part of farming which should be more steadily profitable than the dairy, but, at the same time, not one which demands greater judgment and attention.

Of the three objects of the dairy, namely, milk-selling, cheese- and butter-making, and the raising of young stock direct from the udder, the first is generally the most profitable at the usual price obtained for the milk. The milk-trade, however, can only be carried on within easy reach of a town or a railway, and in recent years the trade has been a good deal cut up by numerous competitors, too many farmers having gone into the business. Prices, therefore, have been depressed, as a general rule, during the time when cheese and butter have been low in value. Still, it is true that milk-selling pays as well or better than anything else in farming, and the hope is that it will improve now that almost all the milk-producing districts have been tapped by railways. Well selected and well managed herds of cows will yield annually from 500 to 700 gallons per cow, and occasionally more, and this even at the low price of 6d. per gallon net will afford remuneration, small though it be. In the course of a paper read before the British Dairy Farmers' Association, on October 7, 1885, "On the necessity for some change in the law in regard to the adulteration of milk," Sir John Bennet Lawes, of Rothamsted, commented on the then somewhat unsatisfactory state of the milk trade in London:—

"It is not merely that the producer receives a very low price for the milk he supplies, for in this he merely shares the fate of all those who obtain their living from the products of the dairy, but because the law which makes it a punishable offence to sell milk that has been adulterated—or what in most cases is milk diluted with water—does not recognise, as a fact, that the quality of the milk from some breeds of cows is so high, that even if it were mixed with a considerable amount of water, it might still be richer than genuine milk, which was the product of other cows fed on a lower description of food. The result is, that the law in regard to the sale of milk, unintentionally gives every encouragement to the sale, not of *pure* unadulterated milk, but of *poor* unadulterated milk.

"It is quite evident that, under such a system, many breeds of cows which produce a high quality of milk, are altogether excluded from the dairy of the farmer, in consequence of his not being able to get a higher price on account of its better quality, while he cannot afford to sell it at the same price as the ordinary milk. As a matter of fact the production of the largest quantity of genuine poor milk is the great secret of success in such a state of affairs.

"At Rothamsted they had been investigating several questions with regard to the production of milk. Now suppose it was found that by a certain combination of foods, a milk of an unusually low percentage of solids could be obtained: would not this be hailed by the milk-selling farmers as a far greater boon than any process by which a much higher percentage could be produced? It may possibly be said such a milk would not be saleable, but this is quite a mistake. It is true the dealers might raise some objection, as although they will not pay for high class milk according to its value, still they like to have it: the public, however, would buy the low percentage milk readily enough. It does not further follow that the milk of poor quality would taste poor: they had lately shown that milk from silage, which from its colour and taste appeared richer than the milk from mangel, was not so rich in reality; and it is a well-known fact that the purest sugars are not those that taste the sweetest. At all events, without anticipating what may be in store in the future, it must be admitted that the principle of offering a premium for the cow which produces the largest quantity of poor milk is one which requires some modification.

"It is quite evident that the weak spot of the present arrangement is the want of some standard or basis on which the trade shall be regulated.

To say that a person shall be punished for selling adulterated milk, and then to leave the definition of what is, and what is not, pure milk to experts and magistrates—who possibly may differ considerably in their views upon the subject—is hardly fair to the producer. It is, however, much more easy to point out the objections to the present system than to suggest a remedy. Assuming it is desirable that a standard should be fixed defining what is pure milk—the question then arises, What is the standard to be? If it were fixed very low, so as to include the poorest milk ever produced, the result would be that all milk would be diluted down to that standard. If on the other hand a high standard were fixed, it would necessarily exclude some very poor, but yet genuine milk.

"Let it be assumed that a certain quality of milk was agreed upon, by producers and experts, as a fair average to represent genuine milk, and that this was declared to be the standard. The result would be that those who were so inclined would be enabled to keep cows yielding a very high quality of milk, as it would no longer be an offence to dilute such milk to the fixed standard. On the other hand, why should not other qualities of milk be sold, provided the seller stated the amount of dilution to which they were subject?

"Another part of the milk trade which required some reform was the large cost incurred in the distribution of the milk, which affects it seriously as a cheap food. The sky blue liquid which used to be sold in London had no pretensions to be called a food, but genuine milk—were it not for the cost of distribution—would be a very cheap food, especially for the young, and although milk is not so well adapted as meat to be the food of grown up people, it is a perfect food for children, and at the same time very much cheaper than meat.

"Supposing the police regulations would admit of such a proceeding, if milk instead of being sent up to its destination in churns, could be conveyed by rail in a tank upon wheels, it might be sold direct to the consumers with very little addition to the cost of production, and it would thus become a staple article of food to the poor. If the demand for milk is to keep pace with the increased supply some such process must be adopted, and the producer and the consumer must be brought together without the intervention of the middle man. Mayfair and Belgravia may still continue to receive their daily supply through the middle man, if such be their wish, but the time is come when an effort should be made to furnish the teeming population of the Metropolis with cheap milk as a portion of their daily sustenance."

The important subject here discussed by Sir John Lawes possesses an equal interest in the United States. There appear, indeed, to be fair grounds for asserting that the time is near when, in the States, all milk will be closely graded when it reaches the market, and every lot sold on its merits, the basis being the total solids. As a matter of fact, this process is already in operation in the discriminating market of Philadelphia, where Mr. G. Abbott adopted it some years ago. In buying milk he makes three grades: milk from registered Jersey and registered Guernsey herds he denominates grade A; that from herds of

Jersey and Guernsey grades is B milk; the product of herds of native or mixed blood is D milk. The A milk must yield not less than 14·50 per cent. of total solids, B milk must give not less than 13·50, and D milk not less than 12·50 per cent. of solids. Mr. Abbott pays ½ cent above the market price of D milk for B milk, and 1½ cents more for A milk. The A and B grades are sold at a corresponding advance in price. Sales of the D grade are confined to wholesale transactions. In supplying milk to retail customers two or three grades in price are as many as can be managed, and the dealer cannot well make more grades in buying than in selling. Mr. Abbott takes milk from upwards of fifty persons, and makes four or five analyses of the milk of each dairy per month. Shippers are urged so to mix their entire milkings that the contents of all their cans will be alike; they use mainly the 40-quart cans, and, as a rule, samples are taken from not more than one can of a shipment. One competent person is constantly engaged in making analyses—not a general chemist, but one specially trained for this work at the Philadelphia College of Pharmacy. The highest total solids ever recorded from the milk of a herd for a single day was 17·12 per cent. from registered Jerseys, and the lowest 11·32 per cent. from common cows. From herds of common or mixed breeds the following yearly averages of total solids were obtained in 1886— 12·60, 12·66, 13·19, and 13·25 per cent. Mr. Abbott believes this system stimulates the production of better milk, and he would not know how to manage his business without grading; he adds that amongst dealers the disposition is increasing to test and grade milk and make a distinction in price according to merit.

The only ways of enlarging the margin of profit to producers of any commodity are to lessen the cost of production and to increase the selling price; to the producer the former is more likely than the latter to be available. The most direct method of reducing the cost of making milk is improvement in the quality of dairy cattle, and what many districts need to produce is not more milk but better milk. In view of the incontrovertible evidence of the influence of breed, much more than feed, upon the quality of milk, producers are exhorted to aim at supplying the top of the market. "Make the market, if need be. Grade your milk. Separate your milk, and encourage the trade in cream."

The making of cheese and butter ranks next in the scale of profit, provided a good article of either is produced. A well-fed cow, of a good breed, will produce, on an average, 200 to 300 lb. of butter in the season; and this, where there is an immediate market for it, together with the value of the skim-milk, either for feeding pigs or raising calves, will pay better than cheese alone. The common calculation is 150 lb.; but that has regard to mixed stock, which affords no certain data.

Mr. Aiton's calculation is, 250 lb. per annum, or 1 lb. of butter from every 10 quarts of milk; but that is for the best milkers of a very superior stock. Although it may be difficult to reach that quantity in any other than a very select dairy, there can be little doubt that, with proper attention to breed and feeding, the Epping average may be maintained.

The average product of *full-milk cheese* or whole-milk cheese in the best English dairies, where the whole milk and cream are used, cannot be estimated at more than four cwt. In Leicestershire and on other deep grazing soils that carry heavy stock, a well-managed cow is reckoned to make from three to five cwt.,[1] besides supporting her calf until it can be weaned; but such cows require fully three acres of the best meadow land, for summer and winter keep, and it is not in the power of every farmer, even if he has the stock, to procure such land to maintain them. In Somersetshire the average is four cwt. and a half;[2] in Essex not so high;[3] and Mr. Marshall states that of all the Midland Counties at something more than three cwt.[4]

The cows of Wiltshire are reckoned to yield from three and a half cwt. to four cwt. of cheese in the year, besides a pound of whey butter per week during the summer season.[5]

Suckling—that is, fattening calves for the butcher—is generally considered the least profitable, as well as the most precarious division of dairy farming, both from the accidents to which calves are liable, and from the more variable price of veal than of butter and cheese. It is, however, the least troublesome; and probably, from the making of butter being combined with it, would be the most advantageous. Supposing a steady weekly demand for butter throughout the year, the most advisable plan might be to keep such a number of cows as would supply that demand during the winter; and in summer, when butter is cheap and veal in request, to apply the extra milk to suckling calves, either for the market or for stock, as may best suit the ulterior views of the farmer. This must, however, depend on the situation of the farm; for that may not always afford an opportunity for the purchase of a succession of calves for suckling, or a market for them when fat; or it may not be adapted for the rearing of stock; and, in such cases, the best application of the skim-milk is either to feed pigs or to raise calves. The usual time required for fattening calves for the butcher has been already stated to be from ten to twelve weeks:[6] perhaps it would be less in summer, when the milk is abundant and the atmosphere genial and warm; but as the calf does not require the entire milk of the cow for some weeks after its birth, she will, for a short period, support two; and two cows, calving at different periods, may be calculated to fatten seven calves in the course of the year.

Compared with *grazing*, every branch of dairy husbandry should be profitable; but the trouble and difficulty of management exceeds the mere feeding of cattle for the shambles. Dairying has also this superiority in other points of considerable importance on farms where the mixed system of tillage and grazing is adopted—it does not require so rich a soil as that for fattening beasts, and it produces food for pigs,

[1] "Leicester Agricultural Survey," pp. 154 and 227. "Cheshire" ditto, p. 271.
[2] "Somerset Agricultural Survey," 3rd edit. p. 251.
[3] "Essex Agricultural Survey," vol. ii. p. 271.
[4] "Rural Economy of the Midland Counties," 2nd edit. vol. i. p. 326.
[5] "Report of 'The Times' Commissioners."
[6] See Book i. chap. vii.

or calves, and thus, by nourishing more animals, creates additional manure and a profitable consumption of the crops on the spot. It has been calculated that the herbage that will add 112 lb. to the weight of an ox will enable a dairy cow to yield 450 gallons of milk, which will be found to exceed the return in meat, after making every fair allowance for the additional expense of management.

Throughout the system of dairy management the vigilant eye of the master should be carefully employed, for the servants will rarely give that minute attention to every particular which is so indispensably necessary in order to ensure success. On this account, it is likely that a dairy-farm of a moderate size—one, for instance, that keeps from ten to twenty cows—will, *if properly managed*, afford a larger proportionate profit than another of greater extent, because, in the former case, the farmer's wife and daughters can more easily superintend, or perhaps perform, a considerable part of the dairy operations themselves; and this will always be better done by them than by hired servants. No branch of husbandry deserves and requires such unremitting attention. Sir John Sinclair very justly remarks, "that if a few spoonfuls of milk are left in the udder of a cow at milking—if any one of the implements used in the dairy is allowed to be tainted by neglect—if the dairy-house is kept dirty or out of order—if the milk is either too hot or too cold at coagulating—if too much or too little rennet is put into the milk—if the whey is not speedily taken off—if too much or too little salt is applied—if the butter is too slowly or too hastily churned—or if other minute attentions are neglected—the milk will be in a great measure lost." If these nice operations occurred only once a month, or once a week, they might be easily guarded against; but, as they *require to be observed through every stage of the process, and almost every hour of the day*, the most vigilant attention must be kept up throughout the whole season. This is not to be expected from hired servants. The wives and daughters of farmers, therefore, having a greater interest in the concern, are more likely to bestow that constant, anxious, and unremitting attention to the dairy, without which it cannot be rendered productive.[1]

CHAPTER VIII.

The Factory System of Dairying—Home and Foreign.

IT will be not only interesting, but it will serve important practical purposes, if we glance, however briefly, at those methods in use in countries other than our own; and at the systems upon which dairy work is there conducted.

[1] Sir John Sinclair on the Husbandry of Scotland, vol. ii. p. 124.

Dairying in the United States of America has long been celebrated for the system with which its details have been carried out, even on farms which we should consider as very small—a system which embraces not merely the arrangement of the buildings, but the fitting up with various contrivances, and the application in working of different kinds of appliances; all calculated to economise time and save labour—both necessities of the situation in which American farmers find themselves, with respect to the difficulty of securing farm servants. As our readers are generally aware, the United States Government have a department which concerns itself wholly with agriculture. This department has done a vast deal of work since the date of its establishment, in the way of appointing commissions of practical men to inquire into various subjects exciting attention amongst, and likely to be of service to, farmers, publishing reports, &c. Convinced of the importance of attending to the production of butter and cheese—with a view to getting rid of the necessity of importing supplies of these from foreign countries—a commission was despatched many years ago to Europe, for the purpose of instituting inquiries in all the countries in which there were districts or localities celebrated for their dairy produce. The report published by the commission, which extended over a very wide range of countries and of districts, was perhaps the most valuable ever issued on any agricultural subject; and this, being widely and wisely distributed through the United States, gave an impetus to dairying which it has never since failed to feel the force of. It gave rise to the systematic working already alluded to, the latest and most successful phase or outcome of which is the "factory system," of which doubtless the majority of our readers have heard, and of which there are now in this country a number of examples.

As usual in the case of all new movements, so in this; the proposal to establish the system met with great opposition. This was not to be wondered at when it came from quarters from which it might with some reason be expected to flow, where an interference with private enterprise was expected and feared; but that it should come from those who had not this excuse, if indeed any, to make, was a matter of surprise to some and disappointment to many. But to some minds all innovations are bad; and it is only when they become successes that they are considered in the more favourable light.

That the factory system is possessed of great advantages, even from a commercial point of view, a fair and candid review of all the circumstances connected with it will lead the majority of inquirers to admit; but that there are difficulties in the way of carrying it out in some districts, and that in a few it is not at all applicable, even its warmest supporters will readily allow. It would indeed be a singularly successful movement which was found to be applicable to all circumstances and all localities.

We cannot, from lack of space, give a full account of all the features connected with the system; nor, if space were at command, would that be necessary, as not coming within the scope of our work. We can only glance at its leading features,—those chiefly which involve points of practical interest. That there will be many such points may readily be conceived; for, in such carefully conducted places as factories, in which every detail is carefully calculated and carried out, experience must have resulted in the deciding of hitherto disputed points of practice, or in the discovery of new ones. Every one at all desirous of excelling, no matter what the circumstances of his dairy, ought to be glad to avail himself of such results of a practice more extended, at all events, if not more carefully carried out, than his own. As that well-known authority, Mr. J. C. Morton, excellently well puts it: "Even on estates already well equipped, the practice of the best and most successful manufacturers ought not to be lightly thought of, either by the landowner or by the farmer." And when we learn that one result of the working of the factory system has been the raising of the market value of cheese—where cheese is the principal if not the only product made in the factories—ten shillings the cwt., as compared with home-made cheese in the same neighbourhood, one may well endorse the statement further made by Mr. Morton, "if the great staple agricultural manufacture of any country can be improved so as to largely increase the value of its annual produce—the fund out of which rent, labour, and the tenant are all paid—it must be pronounced mere sentimental folly to oppose the improvement because estates have been recently equipped at some cost for the former less profitable process."

This is putting the case fairly, candidly, and honestly; and any system which has produced such results should be welcomed by all as a new power adding to the wealth of the community. It is to be noticed, moreover, and that with no small degree ot satisfaction by those interested in the progress of agriculture, that wherever a factory is established the *farming of the neighbourhood* begins to advance, and to rise in the scale of effective working. This, after all, is a natural result. It is an absolute necessity of the factory system that the milk supplied to it by the farmers of the neighbourhood shall be in the best possible condition; other than this will not do for the results they aim at; other than this, therefore, will not be bought. All milk sent to a factory must be clean and fresh, for the impure milk of one farm will vitiate the milk it is mixed with from all the others. The strictest care as to cleanliness is a *sine quâ non*—cleanliness of milk-pails, milk-cans, milk-sieves, and of everything from which harm may be apprehended. Instances of uncleanliness have occurred, at times, at all the factories, and bad cheese has been the result. It follows, therefore, that the manager must keep a watchful eye on all the milk he takes in, so that disobedience to rules may be detected and punished.

In the American factories the practice, as may be supposed, varies considerably; still there is a general principle which runs through them all, so far as the working details, both commercial and farming or dairying are concerned. We do not consider that commercial details come so much within the scope of the present work as those connected with the practical making of butter and cheese, as from these the probability is that our readers may pick up some points of utility in their own practice, or may have suggestions thrown out to them. What we shall give, therefore, on the subject, will be confined merely to such. Those who wish to go fully into the matter will find it detailed in papers in vols. vii. and viii. of the Second Series of the Journal of the Royal Agricultural Society of England; in two papers read before the London Central Farmers' Club, the one by Mr. Henry M. Jenkins, and the other by Mr. John Coleman; together with the papers alluded to and named in other parts of the present chapter.

Taking the papers in the Journal of the Royal Agricultural Society, vol. vii., as our guide, we find the following is the routine of daily work, with a description of the arrangements of the buildings and appliances used. The cooling of the milk is the first and one of the most important parts of the operation. It is generally done by the aid of water obtained from wells or springs, yielding water of as low a temperature as possible. The methods in use for securing such supplies of course depend upon the local features of the springs or wells, and the relation of their level to that of the buildings. In one factory—and amongst the first erected—the springs were so situated that the vats were constructed in such a way as to enclose them. The excavations required were lined with solid masonry, and the depth of the vats, or "pools," as they are termed, was such that the level of the water in them was never higher than that of the floor of the spring-house. Racks were ranged near the bottom of the pools, on which the milk-cans or pails were placed, the water flowing through these racks and above them, to the height of seventeen inches.

Very few instances occur where a spring is available within a factory; and, indeed, a spring so located is not desirable, if only good and cold water can be obtained near to and brought down in pipes. In this way a fall is secured, so that the water has a force that is useful in several ways,—in washing floors and windows, in turning the wheel which actuates the milk-agitator through the night, and so on,—as well as in cooling the milk.

As soon as the evening's milk is received at the cheese factory it is weighed and run into the milk-vats, in which it is cooled and agitated until the morning's milk arrives, when both together are made into cheese. The water runs at the one end into the space between the two shells of the milk-vat, and out of it at the other.

In the case of butter-factories the milk is weighed and put into the milk-pails, which are made of tin, the depth being from 20 to 22 inches, diameter 8 inches. Two pails on the average are required for one cow's milk delivered. The milk is made to reach within four or five inches of the top of the pail, which is immediately placed on the rack in the water-pool, so that the level of water and that of milk are equal. Each pool is arranged to hold about 500 gallons of milk.

The cold water is kept passing through the pool in a continued stream, and if the temperature of the water be properly arranged, the milk should be cleared thoroughly from all animal heat in the space of about an hour. The most suitable temperature of the water is

A A

about 50° Fahr. ; it should not be lower than 48° nor higher than 57°. As showing the diversity of practice—shall we say of opinion ?—the ice process of cooling the water described in connection with the Swedish factories in a subsequent part of this chapter is not approved in the American system now under consideration, the butter made with ice-water being found to be, or supposed to be, more sensitive to heat than that made with cold spring-water.

It is considered of great importance to expose as little as possible of the surface of the milk to the air, in order that the top of the cream may not get dry, this dryness "flecking" the butter, and injuring its flavour. The milk of one day is left in the pools till next morning, giving 24 hours for the morning delivery, and 12 hours for the evening delivery, for the cream to rise. A little funnel-shaped vessel, with a long handle fixed to one side, is used to raise the cream from the pails. As soon as the blue milk level is reached no more cream is taken out.

The cream in autumn and spring is churned sweet as soon as it is taken out of the pails ; in summer it is put into pails and kept in the pools till it has acquired a slightly acid taste, when it is churned. In some factories the cream, as a rule, is churned sourish, the butter-milk going to the cheese-vats with the skim-milk, to be made into " skim-milk cheese."

The churning is generally done by steam power, and the churn preferred in American butter-factories is the Blanchard Churn, as seen in Fig. 72, which is made in different sizes to churn from 30 to 150 gallons. Quick churning, to which we have in another place referred, is not desired, as butter when churned too quickly is injured. A period of from half to three-quarters of an hour is considered the best. The quantity put into the churn at a time is from 60 to 70 quarts, and, with this quantity of cream, from 12 to 16 quarts of water, to dilute and thin the cream and to bring it to a temperature of about 60° F.

Some makers prefer to pass the cream through a sieve—previously diluting it with water—before putting it into the churn. This is done in order to keep back any knotty particles, and to ensure a perfect uniformity in the thickness of the cream. This mode of working is deemed of great importance by some makers of the best qualities of butter, who also prefer thin cream got by putting the milk in deep vessels to the thick " seething cream," obtained by putting it up in shallow ones, which is not evenly churned. The dashers of the churns are arranged to go within an inch of the bottom of the downward stroke, and to rise above the cream in the upward stroke. The temperature of the cream during churning should not be above 60°, and if at the finish the butter-milk should exceed this, the butter will be injured both in flavour and colour. In cold weather the temperature of 62° is the best.

The working up of the butter after it is taken out of the churn is a most important process, much of the quality being dependent upon the way in which this is done. There are various methods of carrying out the process, some dairymaids preferring hand-working, and this used to be the most general way adopted with us. Others prefer to use mechanical butter-workers, these being most highly esteemed in America. Some dairymaids have always warm and often hot, perspiring hands. In such cases there can be no doubt as to what should be done; they should never resort to the hand process. But where they have cool clean hands, it is quite possible that there is nothing which can surpass the delicate manipulative power of the hand working, after some experience is gained. Cold water of the purest quality is essential in washing the butter previous to working. In the American butter factories, they use, in some instances, a water sprinkler, which is simply a miniature watering-can with a fine rose which delivers the water to the butter in finely divided and numerous

Fig. 72.—The Blanchard Churn.

streams. The points to be aimed at in butter-working are the thorough expelling of the butter-milk, and the giving to the butter "firmness of texture and a wax-like appearance when produced."

The salting of the butter is a process which has to be done with care, in order to result in uniform flavour. Formerly, and indeed often now, butter is salted to such a degree that few people can eat it with pleasure; but a purer taste is gradually developing, and a mild saltness is more frequently a favourite than formerly. With us far too little attention—in many cases it may be truly said none—is given to the quality of the salt used. In the American factories the greatest care is taken to have the purest salt; tests are applied to discover the presence of chloride of calcium—if discovered, the salt is discarded, as this substance gives a bitter taste to the butter. Sulphate and carbonate of calcium are also liable to occur. Higgins's

A A 2

dairy salt is so prepared as to eliminate these foreign substances, and is specially suited for use in the dairy.

When the butter is removed from the churn, it is lifted with ladles into wood trays of an oval shape, and the butter-milk is rinsed out with cold water, the ladle being lightly used, so that the water comes off the butter-milk from one end of the tray. This gentle working with the ladle, and washing with the cold water, is repeated till the butter-milk is wholly washed out. The salt is now added—in the proportion of 18 oz. for 22 lb. of butter—and well worked in. It is then allowed to stand till evening—the above processes being of course part of the morning's work—when a second working is given to it, and it is packed for market. This proportion of salt is a high one, and for light salting 5 oz. would be enough for the quantity of butter named.

In the cheese factories, where the whole-milk process is carried out, *whey-butter* is sometimes made. Of course the quality of this is far below that of ordinary butter. Still, by the new process, it is very palatable; and, indeed, so good that, as is stated in the report, experienced dealers having the two kinds offered them without remark being made as to " which is which," the whey-butter has been chosen by them as the better of the two. Whey-butter, as a general rule, rapidly deteriorates, so that it is only fit for immediate use. Nevertheless, if all the water has been worked out, and if the product is potted so closely as to exclude all air, whey butter is found to improve by keeping for some months; a layer, one inch thick, of salt is placed on the top of the jar.

In the American factories the process of making whey-butter is as follows. A vat of copper is employed, 12 feet in length, 3 feet in width, and 20 inches in depth. This is set over an arched furnace, in which wood is the fuel used. The level of the vat and furnace is a little lower than that of the milk-vat, so that the whey can be easily drawn off from the latter. When the vat is filled to its proper height with the whey, " acid " whey is added to the mass in the proportion of 1 gallon to every 50 gallons of sweet whey. If the whey has itself an acid flavour, less " acid " is added in proportion; and if the " acid " itself be not sharp, 1 lb. of salt is added to the above quantity. As soon as the acid is added to the whey, heat is applied to the copper vat, till the temperature is raised to from 170° to 180° F. The cream begins to rise and is skimmed off with a tin scoop, and when wholly removed it is set aside in proper vessels till it cools, and is left to stand for about 24 hours. The cream, thus cooled, is then churned at a temperature of from 56° to 68°, according to that of the weather; and when the butter comes it is taken out and finished off in the usual way. About 20 lb. of butter are thus obtained, on an average, from 500 gallons of whey. The "acid" referred to is obtained by taking whey which is devoid of cream, heating it to the boiling point, and adding 1 gallon of whey which is thoroughly sour to every 10 gallons of boiling whey. The casein and albumin in the mass collect together and can be removed, and the residue is allowed to stand for 24 or 48 hours, according to circumstances, when it is fit to be used as the "acid." Singular to say, the whey left after the butter has been made from it is said to be better adapted for swine-feeding purposes than ordinary whey; this is owing to the sugar of milk being retained longer in the mass without change.

Having described the process of butter-making on the factory system, we shall now glance as briefly as may be at the details of cheese-making, taking as our "model" the example afforded by the Holms factory in Staffordshire which is fully described by Mr. Morton in the paper we have already quoted from, and named below.

We shall in the course of this chapter (page 360) describe the ice method of cooling milk for a butter-making factory, as employed in Sweden. We here describe the "cooling" system as adopted in the Holms factory, near Sheen, full particulars (supplied by Professor J. P. Sheldon) of the working of which will be found in a valuable paper ("Cheese-making in Home Dairies and Factories") by Mr. J. C. Morton, in the Journal of the Royal Agricultural Society of England, vol. xi. second series (1875), page 261.

The cooling vats are of timber, having a milk-holding capacity of 500 gallons, the length of each vat being some 14 feet, width 4 feet, and depth 20 inches. Each vat is lined with tin, or rather has an interior tin vat, the dimensions of which are so much less than those of the main or timber vat, that a hollow space is left at the bottom and sides. The milk is placed in bulk within the tin vat, and cold water, passed into the jacket at one end, passes through the whole length of the space and out at the other. The evening's milk, placed in the vats, is thus surrounded with cold water, and kept exposed to it all the night through. To prevent the cream from rising and also to aërate the milk, and to get rid further of any animal or other odour which may be present in it, wooden stirrers—which sink to a depth of two inches in the milk in the vats—are caused to move to and fro at regular intervals. The stirrers are moved by an ingenious arrangement. The issuing water from the vat, entering one of the buckets of a small water-wheel, fills this till it has weight sufficient to give the wheel half a revolution on its axis. By means of a crank and connecting rod, this motion is communicated to the stirrer.

The milk delivered in the evening, and thus set aside to cool, is reduced in temperature by, and generally before, morning to 60° or 65° F. The morning's milk does not require to be passed through the same long cooling process, but is at once mixed with the evening's milk. When mixed, steam is introduced under vat No. 1—the cold water having been of course previously withdrawn—and the temperature of the milk is raised to about 80° in the summer or warm weather, and 82° in the winter or cold weather. The rennet—which is proportioned to circumstances, but which if in proper condition should be at the rate of half a pint to every 100 gallons of milk—is then put into the vat, and well mixed; and the vats are next covered up with a cloth, to keep in the heat and maintain a uniform temperature. If the rennet is good, it should thicken the milk perceptibly in fifteen minutes, and thoroughly coagulate it in one hour.

119

Coagulation is completed when the curd will break cleanly over the finger; it is then cut by the curd-cutter, the cutting being done slowly from one end of the vat to the other, and repeated till the whole mass is cleanly cut, not bruised and broken. The curd thus cut is allowed to remain quiescent for a few minutes, till the whey rises and covers the surface, when the curd-cutter is again passed through the mass, but in a direction at right angles to the previous cut; thus leaving the curd in the form of cubical blocks, or rather parallelopidedons, say half an inch square on the side. The whey is then allowed to escape from the vat, and the curd slowly gravitates to the lower part of the vat. Allowed to remain quiet thus for a short time, the mass is gently turned over by the hands, and then it is cut into square blocks by the equally gentle use of a knife. These and all succeeding movements of the curd must be done with the utmost care—tenderness, as Mr. Morton well expresses it,—for it is essential to retain as much as possible of the fatty matter in the curd, and to allow the minimum only to pass off along with the whey. A little steam is now turned into the empty water space, and as soon as the temperature is slightly raised by it, the curd, acquiring more firmness, can be manipulated a trifle more freely, and turned about faster; this brings out the whey more quickly, and correspondingly reduces the bulk of the curd. More steam is turned on, and the curd is stirred more quickly than before, to prevent any over-heating of the mass at the bottom of the vat. The whey being by this time nearly wholly expelled from the curd, the latter has become hard and tough, and the curd-rake is freely used to keep its particles in motion. When a temperature of 90° F. is reached, the steam is turned off, and the curd is kept stirred till the bottom of the vat has gradually cooled. It is then allowed to remain quiescent for about ten minutes, when the steam is again turned on at full pressure, and the curd is kept in continual motion. When 100° F. is reached, the steam is for the last time turned off, and the curd kept worked till the vat has gradually cooled down.

The curd is now left till the "souring process" is completed, the time for effecting which is dependent upon circumstances, and is decided by the experience of the manager. A test sometimes employed is that of taking out a piece of curd and applying it to hot (not red-hot) iron; if it draws out into fine threads of about an inch long the curd is in good condition. Litmus paper, however, affords the safer test. The whey which has collected is run off from the vat by a syphon pipe; and still further to get rid of what remains, the curd is gathered up towards each side of the vat, till a space is left up the middle into which the whey runs. The whey, indeed, is run off before it has become acid, and the requisite acidity is developed in the curd alone, which is kept warm in the vat with that object. The curd, now adhering in a mass, is cut into pieces, and turned over and over till all the whey is expressed. It is then taken out of the vat, put into the curd-mill, and reduced to something like currants and raisins in size; to this salt is added—at the rate of 2 lb. to every 1,000 lb. of the milk, autumn-made cheese having a higher proportion of salt, about $2\frac{1}{2}$ lb.

The salted curd is next vatted, and subjected to the action of screw presses till the last portion of whey is expressed. In these presses it remains till next morning, when it is taken out, and conveyed to the lower curing-room and weighed, has some tissue paper attached to the flat sides of what is now a formed cheese, and is placed on the cheese-shelves to cure and ripen. Here it is turned each day for a few days, when it goes to the upper curing-room, on the shelves of which it is turned every other day.

In some cases the curd is placed in what is technically termed a "dry-vat," so soon as the whey has been taken off it. This vat is generally a good deal smaller than those in which the milk is coagulated, and it is provided with a false perforated bottom, on which a cloth strainer is placed to facilitate the passing out of the whey.

The round form of the cheese, so well known, is obtained by using strong circular hoops of wood or of galvanized iron, the diameter being on the average 15 inches and the depth 5 or 6 inches. The hoops are placed on a movable board at the bottom of the press. The curd is filled into the hoop, and carries a cloth with it to the bottom of the hoop, the cloth being of course first placed over the hoop. When the hoop is filled with curd another cloth is placed over it, then a small board, and the whole is slipped under the press. The pressure is applied slightly at first, and then gradually increased till it reaches that of four tons. When the curd is solidified, it is ready to receive the permanent bandage, which is of stout calico. This bandage is so arranged that it covers the round edge of the cheese, with an overlay at top and bottom. The cheese is then returned to the hoop, and again subjected to pressure for about 18 to 24 hours, when the cheese is taken out and carried to the curing-room. In the factory in which the process now being described is carried on, the cheese, after being removed from the hoop, is rubbed over with whey-butter, for two or three days. This is done to prevent the cracking of the outer skin or rind of the cheese. The uniform temperature of the curing-room is of great importance; this for the first six weeks should be about 70° F., when it should be gradually reduced to about 65°, at which it should remain until the cheeses are sold.

In the American factories where the butter is made from sweet cream—not soured or lappered,—the skim-milk, being also sweet, is available for the making of skim-milk cheese, which forms part generally of the operations of butter-making factories. In making the skim-milk cheese, the milk is set in the vat at a temperature of 82° F., and sufficient rennet is added to coagulate the mass in 40 or 60 minutes. The process throughout is very similar to that we have above described. Here it is that the employment of animal fats found an opening, to supply the want of the cream removed,—hence we have that abomination of the dairy, oleomargarine cheese. This product is also called "filled" cheese, because after the removal from the milk of the natural butter-fat, the place of the latter is, as it were, "filled" up by another fat.

Butter factories have been established in Sweden with marked success. As the farms are as a rule small, and the herds of cows kept equally so, the direct delivery of the milk to the company or factory is not available. We have alluded to this difficulty as existing in this country, and as being one urged by many farmers against the factory system. In Sweden, as also in America, at some of the factories this difficulty is got over by purchasing the cream only,—this is known as the "cream-gathering" system,—leaving the skim-milk in the farmer's hands to be dealt with as his circumstances may dictate, as in the making of skim-milk cheese, the feeding of calves, pigs, &c. In order to facilitate the collection and disposal of the milk, even of the smallest farms, small "milk receiving houses" are erected at various points. These are fitted up with the appliances necessary to cool the milk, receive and retain the cream, with washing or scalding-room to cleanse the vessels—a cheese-making room if necessary—and accommodation for the dairymaid. "This system has," says M. Juhlin-Dannfelt, in the "Journal of the Royal Agricultural Society of England," vol. viii., second series (1872), "decidedly promoted the further development of the factory system, and at the same time opened the way to a useful and profitable branch of industry to those who occupy themselves with collecting pure milk from the smaller farmers, whose produce is too limited to allow the cream obtained from it to be treated in the manner which will make it saleable to the dairy company, or from such larger producers of milk as do not care to take the pains necessary for obtaining the cream, or for the further preparation of the skim-milk."

The importance of cooling the milk as soon as possible after it comes from the cow is fully recognised by the Swedish companies, and means are furnished in each of the "milk receiving houses" for having this process carried out quickly and efficiently. A room is provided with a cold-water cistern, from which is drawn the supply necessary to fill the cooling vats in which the milk-pails are placed. If water cold enough is not obtainable, ice is used to reduce its temperature; and, indeed, so much more satisfactory are the results of ice-cooling, that it is now generally used.

As bearing upon various practical points connected with butter-making, &c., we here give a brief *résumé* of the facts detailed in the paper above alluded to. The ice used to cool the water in the supply cistern is broken into pieces some three or four inches square, as the cooling action is found much increased by this. The cooling vats in which the milk-pails are placed are about 9 feet long and 3 feet wide, with a depth of about 2 feet. A false grated bottom is provided to each cistern, and upon this the milk-pails are placed. A vat of the above dimensions is capable of cooling about 115 imperial gallons of milk. The pails were originally about 24 inches deep, and about 18 inches in diameter; but as it has been found that the quicker the milk is cooled the more completely is the cream separated from it, the size has been reduced to 20 inches in depth and about 9 inches in diameter, so that each holds about $3\frac{1}{2}$ gallons. The depth of the iced water in the vats should be such that in summer time it is equal to the height of the milk in the pails—that is the level of each should be coincident. The cream in this arrangement, as it rises, is kept cool; but in the winter season the level of the surface of milk in the pails should be above that of the water in the vat by some inches. The proportion of the ice used in cooling the water to the milk to be cooled varies, but on an average the quantities of each are about equal. By careful management the ice has been reduced to one-third; thus, in the Central Company at Stockholm, 1,500 cwt. of ice were used to cool the milk necessary to produce 2,500 cwt. of butter.

On the milk being delivered at the receiving houses, it is measured, and a small sample put into a graduated glass cylinder, which is left for cream setting, so that an idea may be obtained of its quality. The milk is then strained into the pails, and these are placed in the vats with about three inches interval between them. The temperature of the milk-room—the ice-water in the vats, and that of the milk itself—deciding the rising of the cream, the time taken by the latter varies. On an average the milk will be ready for skimming in about 10 to 12 hours, with a temperature of 35° of the ice-water, but more and better cream is got by allowing it to stand for 18 or 24 hours. The temperature of the milk-room should be as low as possible in summer, never below 50° in winter. The sooner the cream is churned the better, is the experience of the Swedish factories—or, as their reporter puts it, "the fresher and absolutely sweeter the cream is, the better will the butter be." As we have seen from experience in this country, opinion differs on this. On arrival at the factory the cream is put into the ice-water vats at once to keep it cool and sweet. The average results of working may be stated thus: 2·65 gallons of milk yield 0·44, or nearly half a gallon of cream; this churned gives 0·93, or nearly one English pound of butter. The temperature of the cream found best for churning varies with the quality of the cream, the temperature of the churning-room, &c. There is one great advantage obtained by using iced-water in place of cold well-water—even if that can be obtained at a temperature sufficiently low,—and this is that while the well-water in the cooling vats requires to be constantly changed, the ice-water need not be changed oftener than a few times in the year. The surplus water from the melting of the ice is carried off by a small pipe, the orifice of which is near the upper edge of the vat.

The ice is not stored in a regularly-built house, but simply heaped up in a pyramidal form, in the open air generally, but sometimes in a shed, and covered with saw-dust, tanners' bark, or other good non-conducting material. Great care is taken to keep down the number and size of the spaces or interstices between the blocks of ice, and these where they exist are carefully filled up with sawdust. When ice is removed from the heap it is taken from the top, working downwards, and the spaces made by the removal are carefully filled up with saw-dust. The lowest layer, or the bottom blocks, rest upon a layer of the non-conducting material at least a foot in depth.

The dairy factory system has made remarkable strides in Denmark, a circumstance that is chiefly due to the fact that a ready market was found in England for Danish butter. In a paper on "Dairying in Denmark," which was published in the Journal of the Royal Agricultural Society (vol. xix., second series, 1883), Mr. H. M. Jenkins described the widely different details of the process of manufacture in the cases of fresh butter and keeping butter respectively. The former he terms a wet process, the latter a dry one. To make the best fresh butter hard pressing seems unnecessary, but to make the best keeping butter it is deemed to be essential. Whether, however, intended for immediate or for future consumption, butter is nearly always made in Denmark from

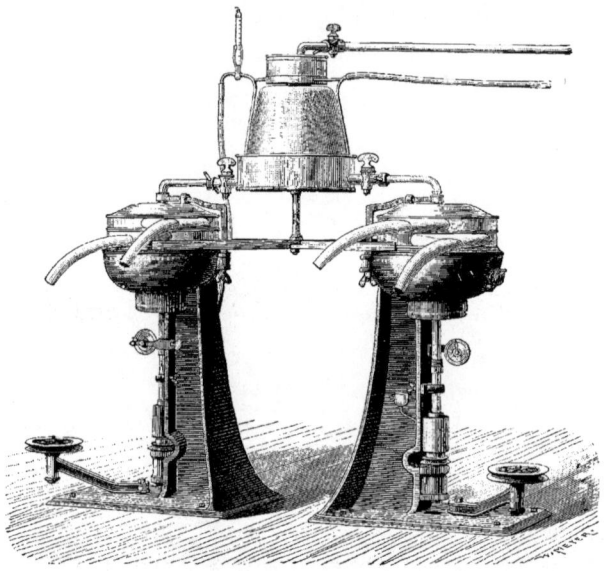

Fig. 73.—Two Laval Separators with Milk-Warmer in Action.

cream which has been taken from sweet milk, but which has afterwards been artificially soured.

Within recent years cheese- and butter-factories have multiplied in number pretty quickly in England, Scotland, and Ireland. When the factory system was first introduced, the chief reason why it did not spread rapidly was the expanding milk-trade of the day; now, however, the milk-trade is the chief reason why it is adopted, because the two are found to work well together. The factory is the receiving-house; and whatever milk there is, beyond the requirement of the trade for the day, is made into cheese, or butter, without loss or inconvenience. In 1870, the late Lord Vernon offered to put up a cheese-factory for the use of his tenantry; but they declined it because, being close to a railway, they were enabled to devote their energies to the milk-trade, which, in the form it has since assumed, was then a comparatively new thing. In 1884, the present Lord Vernon, who treads so well in his father's steps in agricultural reforms, erected a combined butter- and cheese-factory for the benefit of his tenantry, and others around, who had discovered that such an institution was needed not only for the manufacture of butter and cheese, one or both, as the case might be, but also to accommodate the vicissitudes of the milk-trade,

Fig. 74.—Steam-Power Factory Churn, the "Anglo-Hibernia."

in which competition had by that time become very keen and general. The factory on Lord Vernon's estate at Sudbury, in Derbyshire, has had already a long career of gratifying success, and the Sudbury butter commands a high price,—far above the average for fresh butter. As we have said before, however, butter-factories are very differently conducted from what they were at first, for the centrifugal cream-separator has completely displaced the ice-water system of setting milk to cream, and indeed all other systems too. We have previously spoken of separators (p. 261), and have given illustrations of hand and power machines; in Fig. 73 we give a woodcut of a machine for steam or water power— a double or twin machine, indeed, capable of separating 150 gallons an hour, for which the Dairy Supply Company are the sole agents in this country.

By the aid of one or more separators, the milk is deprived of its cream shortly after arriving at the factory; the skim-milk is sold as far as possible, made into cheese, or fed to pigs and calves. The cream, being so greatly reduced in bulk as compared with the milk from which it was obtained, is easily taken care of until it is ripe enough to be churned. Fig. 74 affords a view of Bradford's Factory Churn, made in sizes to churn from 50 to 200 gallons of cream, yielding from 150 to 600 lb. of butter.

Where, on a large scale, as in a butter-factory, cream is obtained and churned by the aid of steam, it is obvious that a steam-power butter-worker will also be most appropriately employed. Indeed, where six

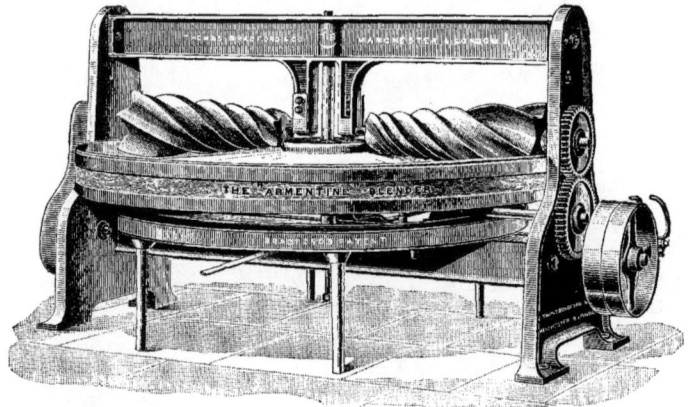

Fig. 75.—Butter-Worker and Blender.

or seven hundred pounds of butter are churned in one operation, a butter-worker of great capacity becomes a necessity, for to manipulate such masses of the golden product by hand-labour alone would be a task requiring a good many pairs of hands. The large machines suitable for butter-factories consist, as will be seen in the illustration, Fig. 75, of a circular table revolving horizontally, on which the butter is placed, and of fluted rollers, which, in a fixed position, manipulate all the butter successively, as the table revolves. The attendant turns up the fluted butter, so soon as it has passed under the rollers, and it is again and again flattened out, until the moisture has been extracted, the salt worked in, and the butter consolidated into a mass, firm and compact in texture. This machine is called a "blender," because it is most effectual in mixing together firkins of butter of different shades of colour, until it is of uniform character throughout. Fig. 76 affords the reader a view of a steam-power butter factory.

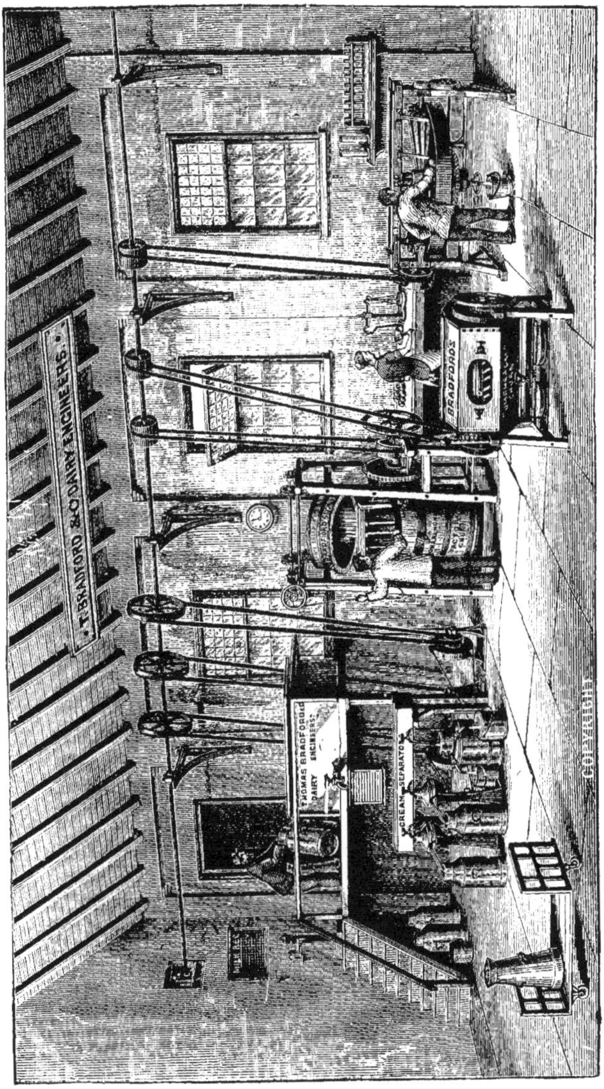

Fig. 76.—Plan of Steam-Power Butter Factory.

In America, and on the Continent of Europe, where the dairy factory or large farm dairy system is in vogue, *pig-keeping* generally forms part of it. To some, but perhaps not to the same, extent, this is the case in this country. The three products of the dairy useful for pig-feeding are whey, skim-milk, and butter-milk. In the neighbourhood of large towns, there is no difficulty in disposing of the two latter products.

But most cheese- and butter-factories are naturally situated in the rural districts, and for obvious reasons; and, being thus far from towns, the bye-products must be used near the farm or factory. Like many other agricultural questions, that of, Do pigs pay? is answered both affirmatively and negatively. In the paper by Mr. Morton, previously alluded to, the question of pig-feeding at dairies is freely discussed, and the discrepancy of opinion just noted is there exemplified; for while one farmer informed Mr. Morton that he would gladly give his whey away for nothing to anyone who would come to his dairy to take it, another valued it so much that he estimated that it was worth thirty-five shillings per cow a year to him; and another attaches so high a value to the whey, "that he justifies by its use the expenditure of 300*l.* a year for the purchase of feeding materials to be used along with it."

There is for the skim-milk a new outlet, when used for feeding purposes,—or rather we should say there are two new outlets,—first for the feeding of calves, secondly for the feeding or fattening of cattle, a recently-introduced system, but which seems to be more applicable to dairy cows, as it adds considerably to the yield of their milk. In the paper on Swedish butter factories, from which we have already culled some interesting facts on dairy management, there is a notice of the method of using the skim-milk for the feeding of calves, which lays before the reader some suggestive facts. From this notice it would appear that the use of skim-milk for this purpose has been eminently successful; the only drawback to it being that it gives a darkness to the flesh. This is, however, avoided by feeding the calves —when killed for veal—on sweet milk for the last fortnight before being killed. Separated milk—that which comes from the centrifugal separator—is profitably used in the making of bread.

Skim-milk, indeed, is valuable in a high degree for various kinds of animals, not calves and pigs and cows only, but horses too. For colts and fillies recently weaned, it is worthy of the highest possible recommendation in reference to the formation of bone and sinew, and to the steady growth of the animal in all respects.

The practice of giving much corn of any kind to young equine animals is known to be highly injudicious; but skim-milk is perfectly safe to use, for it will cause the animal to thrive well, and at the same time will do no harm whatever to the constitution, will form no humours, and develop no unsoundness.

Made in the USA
San Bernardino, CA
10 April 2018